Benjamin Peirce

Linear Associative Algebra

Benjamin Peirce

Linear Associative Algebra

ISBN/EAN: 9783743317987

Manufactured in Europe, USA, Canada, Australia, Japa

Cover: Foto ©berggeist007 / pixelio.de

Manufactured and distributed by brebook publishing software (www.brebook.com)

Benjamin Peirce

Linear Associative Algebra

Linear Associative Algebra.

By BENJAMIN PEIRCE, LL. D.

LATE PERKINS PROFESSOR OF ASTRONOMY AND MATHEMATICS IN HARVARD UNIVERSITY
AND SUPERINTENDENT OF THE UNITED STATES COAST SURVEY.

New Edition, with Addenda and Notes, by C. S. PEIRCE, Son of the Author.

[*Extracted from The American Journal of Mathematics.*]

NEW YORK : D. VAN NOSTRAND, PUBLISHER.
1882.

PREFACE.

Lithographed copies of this book were distributed by Professor Peirce among his friends in 1870. The present issue consists of separate copies extracted from *The American Journal of Mathematics*, where the work has at length been published.*
The body of the text has been printed directly from the lithograph with only slight verbal changes. Appended to it will be found a reprint of a paper by Professor Peirce, dated 1875, and two brief contributions by the editor. The foot-notes contain transformations of several of the algebras, as well as what appeared necessary in order to complete the analysis in the text at a few points. A relative form is also given for each algebra; for the rule in *Addendum* II. by which such forms may be immediately written down, was unknown until the printing was approaching completion.

The original edition was prefaced by this dedication :

To My Friends.

This work has been the pleasantest mathematical effort of my life. In no other have I seemed to myself to have received so full a reward for my mental labor in the novelty and breadth of the results. I presume that to the uninitiated the formulae will appear cold and cheerless; but let it be remembered that, like other mathematical formulae, they find their origin in the divine source of all geometry. Whether I shall have the satisfaction of taking part in their exposition, or whether that will remain for some more profound expositor, will be seen in the future.

B. P.

---

* To page n of this issue corresponds page $n+96$ of Vol. IV. of *The Journal*.

LINEAR ASSOCIATIVE ALGEBRA.

1. Mathematics is the science which draws necessary conclusions.

This definition of mathematics is wider than that which is ordinarily given, and by which its range is limited to quantitative research. The ordinary definition, like those of other sciences, is objective; whereas this is subjective. Recent investigations, of which quaternions is the most noteworthy instance, make it manifest that the old definition is too restricted. The sphere of mathematics is here extended, in accordance with the derivation of its name, to all demonstrative research, so as to include all knowledge strictly capable of dogmatic teaching. Mathematics is not the discoverer of laws, for it is not induction; neither is it the framer of theories, for it is not hypothesis; but it is the judge over both, and it is the arbiter to which each must refer its claims; and neither law can rule nor theory explain without the sanction of mathematics. It deduces from a law all its consequences, and develops them into the suitable form for comparison with observation, and thereby measures the strength of the argument from observation in favor of a proposed law or of a proposed form of application of a law.

Mathematics, under this definition, belongs to every enquiry, moral as well as physical. Even the rules of logic, by which it is rigidly bound, could not be deduced without its aid. The laws of argument admit of simple statement, but they must be curiously transposed before they can be applied to the living speech and verified by observation. In its pure and simple form the syllogism cannot be directly compared with all experience, or it would not have required an

Aristotle to discover it. It must be transmuted into all the possible shapes in which reasoning loves to clothe itself. The transmutation is the mathematical process in the establishment of the law. Of some sciences, it is so large a portion that they have been quite abandoned to the mathematician,—which may not have been altogether to the advantage of philosophy. Such is the case with geometry and analytic mechanics. But in many other sciences, as in all those of mental philosophy and most of the branches of natural history, the deductions are so immediate and of such simple construction, that it is of no practical use to separate the mathematical portion and subject it to isolated discussion.

2. The branches of mathematics are as various as the sciences to which they belong, and each subject of physical enquiry has its appropriate mathematics. In every form of material manifestation, there is a corresponding form of human thought, so that the human mind is as wide in its range of thought as the physical universe in which it thinks. The two are wonderfully matched. But where there is a great diversity of physical appearance, there is often a close resemblance in the processes of deduction. It is important, therefore, to separate the intellectual work from the external form. Symbols must be adopted which may serve for the embodiment of forms of argument, without being trammeled by the conditions of external representation or special interpretation. The words of common language are usually unfit for this purpose, so that other symbols must be adopted, and mathematics treated by such symbols is called *algebra*. Algebra, then, is formal mathematics.

3. All relations are either qualitative or quantitative. Qualitative relations can be considered by themselves without regard to quantity. The algebra of such enquiries may be called logical algebra, of which a fine example is given by Boole.

Quantitative relations may also be considered by themselves without regard to quality. They belong to arithmetic, and the corresponding algebra is the common or arithmetical algebra.

In all other algebras both relations must be combined, and the algebra must conform to the character of the relations.

4. The symbols of an algebra, with the laws of combination, constitute its *language*; the methods of using the symbols in the drawing of inferences is its *art*; and their interpretation is its *scientific application*. This three-fold analysis of algebra is adopted from President Hill, of Harvard University, and is made the basis of a division into books.

PEIRCE: *Linear Associative Algebra.* 3

BOOK I.*

THE LANGUAGE OF ALGEBRA.

5. The language of algebra has its alphabet, vocabulary, and grammar.

6. The symbols of algebra are of two kinds: one class represent its fundamental conceptions and may be called its *letters*, and the other represent the relations or modes of combination of the letters and are called *the signs*.

7. The *alphabet* of an algebra consists of its letters; the *vocabulary* defines its signs and the elementary combinations of its letters; and the *grammar* gives the rules of composition by which the letters and signs are united into a complete and consistent system.

The Alphabet.

8. Algebras may be distinguished from each other by the number of their independent fundamental conceptions, or of the letters of their alphabet. Thus an algebra which has only one letter in its alphabet is a *single* algebra; one which has two letters is a *double* algebra; one of three letters a *triple* algebra; one of four letters a *quadruple* algebra, and so on.

This artificial division of the algebras is cold and uninstructive like the artificial Linnean system of botany. But it is useful in a preliminary investigation of algebras, until a sufficient variety is obtained to afford the material for a natural classification.

Each fundamental conception may be called a *unit*; and thus each unit has its corresponding letter, and the two words, unit and letter, may often be used indiscriminately in place of each other, when it cannot cause confusion.

9. The present investigation, not usually extending beyond the sextuple algebra, limits the demand of the algebra for the most part to six letters; and the six letters, i, j, k, l, m and n, will be restricted to this use except in special cases.

10. *For any given letter another may be substituted*, provided a new letter represents a combination of the original letters of which the replaced letter is a necessary component.

For example, any combination of two letters, which is entirely dependent for its value upon both of its components, such as their sum, difference, or product, may be substituted for either of them.

---

* Only this book was ever written. [C. S. P.]

This *principle of the substitution of letters* is radically important, and is a leading element of originality in the present investigation; and without it, such an investigation would have been impossible. It enables the geometer to analyse an algebra, reduce it to its simplest and characteristic forms, and compare it with other algebras. It involves in its principle a corresponding substitution of units of which it is in reality the formal representative.

There is, however, no danger in working with the symbols, irrespective of the ideas attached to them, and the consideration of the change of the original conceptions may be safely reserved for the *book of interpretation*.

11. In making the substitution of letters, the original letter will be preserved with the distinction of a subscript number.

Thus, for the letter i there may successively be substituted i_1, i_2, i_3, etc. In the final forms, the subscript numbers can be omitted, and they may be omitted at any period of the investigation, when it will not produce confusion.

It will be practically found that these subscript numbers need scarcely ever be written. They pass through the mind, as a sure ideal protection from erroneous substitution, but disappear from the writing with the same facility with which those evanescent chemical compounds, which are essential to the theory of transformation, escape the eye of the observer.

12. A *pure* algebra is one in which every letter is connected by some indissoluble relation with every other letter.

13. When the letters of an algebra can be separated into two groups, which are mutually independent, it is a *mixed algebra*. It is mixed even when there are letters common to the two groups, provided those which are not common to the two groups are mutually independent. Were an algebra employed for the simultaneous discussion of distinct classes of phenomena, such as those of sound and light, and were the peculiar units of each class to have their appropriate letters, but were there no recognized dependence of the phenomena upon each other, so that the phenomena of each class might have been submitted to independent research, the one algebra would be actually a mixture of two algebras, one appropriate to sound, the other to light.

It may be farther observed that when, in such a case as this, the component algebras are identical in form, they are reduced to the case of one algebra with two diverse interpretations.

The Vocabulary.

14. Letters which are not appropriated to the alphabet of the algebra *
may be used in any convenient sense. But it is well to employ *the small letters*
for expressions of common algebra, and *the capital letters* for those of the algebra
under discussion.

There must, however, be exceptions to this notation; thus the letter D will
denote the derivative of an expression to which it is applied, and Σ the summation of cognate expressions, and other exceptions will be mentioned as they
occur. Greek letters will generally be reserved for angular and functional
notation.

15. The three symbols J, Θ, and $\mathfrak{G}$ will be adopted with the signification

$$J = \sqrt{-1}$$

Θ = the ratio of circumference to diameter of circle = 3.1415926536
$\mathfrak{G}$ = the base of Naperian logarithms = 2.7182818285.

which gives the mysterious formula

$$J^{-J} = \sqrt{\mathfrak{G}^\Theta} = 4.810477381.$$

16. All the signs of common algebra will be adopted; but any signification
will be permitted them which is not inconsistent with their use in common
algebra; so that, if by any process an expression to which they refer is reduced
to one of common algebra, they must resume their ordinary signification.

17. The sign $=$, which is called that of equality, is used in its ordinary sense
to denote that the two expressions which it separates are the same whole,
although they represent different combinations of parts.

18. The signs $>$ and $<$ which are those of inequality, and denote "more
than" or "less than" in quantity, will be used to denote the relations of a whole
to its part, so that the symbol which denotes the part shall be at the vertex of
the angle, and that which denotes the whole at its opening. This involves the
proposition that the smaller of the quantities is included in the class expressed
by the larger. Thus

$$B < A \text{ or } A > B$$

denotes that A is a whole of which B is a part, so that all B is A.†

---

* See § 9.
† The formula in the text implies, also, that some A is not B. [C. S. P.]

If the usual algebra had originated in qualitative, instead of quantitative, investigations, the use of the symbols might easily have been reversed; for it seems that all conceptions involved in A must also be involved in B, so that B is more than A in the sense that it involves more ideas.

The combined expression
$$B > C < A$$
denotes that there are quantities expressed by C which belong to the class A and also to the class B. It implies, therefore, that some B is A and that some A is B.* The intermediate C might be omitted if this were the only proposition intended to be expressed, and we might write
$$B ><A.$$
In like manner the combined expression
$$B < C > A$$
denotes that there is a class which includes both A and B,† which proposition might be written
$$B <> A.$$

19. A vertical mark drawn through either of the preceding signs reverses its signification. Thus
$$A \neq B$$
denotes that B and A are essentially different wholes;
$$A \not> B \text{ or } B \not< A$$
denotes that all B is not A, ‡ so that if they have only quantitative relations, they must bear to each other the relation of
$$A = B \text{ or } A < B.$$

20. The sign $+$ is called *plus* in common algebra and denotes *addition*. It may be retained with the same name, and the process which it indicates may be called addition. In the simplest cases it expresses a mere mixture, in which

---

* This, of course, supposes that C does not vanish. [C. S. P.]
† The universe will be such a class unless A or B is the universe. [C. S. P.]
‡ The general interpretation is rather that either A and B are identical or that some B is not A. [C. S. P.]

PEIRCE: *Linear Associative Algebra.* 7

the elements preserve their mutual independence. If the elements cannot be mixed without mutual action and a consequent change of constitution, the mere union is still expressed by the sign of addition, although some other symbol is required to express the character of the mixture as a peculiar compound having properties different from its elements. It is obvious from the simplicity of the union recognized in this sign, that the order of the admixture of the elements cannot affect it; so that it may be assumed that

$$A + B = B + A$$
and
$$(A + B) + C = A + (B + C) = A + B + C.$$

21. The sign — is called *minus* in common algebra, and denotes *subtraction*. Retaining the same name, the process is to be regarded as the reverse of addition; so that if an expression is first added and then subtracted, or the reverse, it disappears from the result; or, in algebraic phrase, it is *canceled*. This gives the equations

$$A + B - B = A - B + B = A$$
and
$$B - B = 0.$$

The sign minus is called the negative sign in ordinary algebra, and any term preceded by it may be united with it, and the combination may be called a *negative term*. This use will be adopted into all the algebras, with the provision that the derivation of the word negative must not transmit its interpretation.

22. The sign × may be adopted from ordinary algebra with the name of the sign of *multiplication*, but without reference to the meaning of the process. The result of multiplication is to be called the *product*. The terms which are combined by the sign of multiplication may be called *factors*; the factor which precedes the sign being distinguished as the *multiplier*, and that which follows it being the *multiplicand*. The words multiplier, multiplicand, and product, may also be conveniently replaced by the terms adopted by Hamilton, of *facient*, *faciend*, and *factum*. Thus the equation of the product is

multiplier × multiplicand = product; *or* facient × faciend = factum.

When letters are used, the sign of multiplication can be *omitted* as in ordinary algebra.

23. When an expression used as a factor in certain combinations gives a product which vanishes, it may be called in those combinations a *nilfactor*. Where as the multiplier it produces vanishing products it is *nilfacient*, but where it is the multiplicand of such a product it is *nilfaciend*.

24. When an expression used as a factor in certain combinations overpowers the other factors and is itself the product, it may be called an *idemfactor*. When in the production of such a result it is the multiplier, it is *idemfacient*, but when it is the multiplicand it is *idemfaciend*.

25. When an expression raised to the square or any higher power vanishes, it may be called *nilpotent*; but when, raised to a square or higher power, it gives itself as the result, it may be called *idempotent*.

The defining equation of nilpotent and idempotent expressions are respectively $A^n = 0$, and $A^n = A$; but with reference to idempotent expressions, it will always be assumed that they are of the form

$$A^2 = A,$$

unless it be otherwise distinctly stated.

26. *Division* is the reverse of multiplication, by which its results are verified. It is the process for obtaining one of the factors of a given product when the other factor is given. It is important to distinguish the position of the given factor, whether it is facient or faciend. This can be readily indicated by combining the sign of multiplication, and placing it before or after the given factor just as it stands in the product. Thus when the multiplier is the given factor, the correct equation of division is

$$\text{quotient} = \frac{\text{dividend}}{\text{divisor} \times}$$

and the equation of verification is

$$\text{divisor} \times \text{quotient} = \text{dividend}.$$

But when the multiplicand is the given factor, the equation of division is

$$\text{quotient} = \frac{\text{dividend}}{\times \text{divisor}}$$

and the equation of verification is

$$\text{quotient} \times \text{divisor} = \text{dividend}.$$

27. Exponents may be introduced just as in ordinary algebra, and they may even be permitted to assume the forms of the algebra under discussion.

There seems to be no necessary restriction to giving them even a wider range and introducing into one algebra the exponents from another. Other signs will be defined when they are needed.

The definition of the fundamental operations is an essential part of the vocabulary, but as it is subject to the rules of grammar which may be adopted, it must be reserved for special investigation in the different algebras.

The Grammar.

28. Quantity enters as a form of thought into every inference. It is always implied in the syllogism. It may not, however, be the direct object of inquiry; so that there may be logical and chemical algebras into which it only enters accidentally, agreeably to § 1. But where it is recognized, it should be received in its most general form and in all its variety. The algebra is otherwise unnecessarily restricted, and cannot enjoy the benefit of the most fruitful forms of philosophical discussion. But while it is thus introduced as a part of the formal algebra, it is *subject to every degree and kind of limitation in its interpretation.*

The free introduction of quantity into an algebra does not even involve the reception of its unit as one of the independent units of the algebra. But it is probable that without such a unit, no algebra is adapted to useful investigation. It is so admitted into quaternions, and its admission seems to have misled some philosophers into the opinion that quaternions is a triple and not a quadruple algebra. This will be the more evident from the form in which quaternions first present themselves in the present investigation, and in which the unit of quantity is not distinctly recognizable without a transmutation of the form.*

29. The introduction of quantity into an algebra naturally carries with it, not only the notation of ordinary algebra, but likewise many of the rules to which it is subject. Thus, when a quantity is a factor of a product, it has the

---

* Hamilton's total exclusion of the imaginary of ordinary algebra from the calculus as well as from the interpretation of quaternions will not probably be accepted in the future development of this algebra. It evinces the resources of his genius that he was able to accomplish his investigations under these trammels. But like the restrictions of the ancient geometry, they are inconsistent with the generalizations and broad philosophy of modern science. With the restoration of the ordinary imaginary, quaternions becomes Hamilton's biquaternions. From this point of view, all the algebras of this research would be called bi-algebras. But with the ordinary imaginary is involved a vast power of research, and the distinction of names should correspond: and the algebra which loses it should have its restricted nature indicated by such a name as that of a *semi-algebra*.

same influence whether it be facient or faciend, so that with the notation of § 14, there is the equation
$$Aa = aA,$$
and in such a product, the quantity a may be called the *coefficient*.

In like manner, terms which only differ in their coefficients, may be added by adding their coefficients; thus,
$$(a \pm b)A = aA \pm bA = Aa \pm Ab = A(a \pm b).$$

30. The exceeding simplicity of the conception of an equation involves the identity of the equations
$$A = B \text{ and } B = A$$
and the substitution of B for A in every expression, so that
$$MA \pm C = MB \pm C,$$
or that, *the members of an equation may be mutually transposed or simultaneously increased or decreased or multiplied or divided by equal expressions.*

31. How far the principle of § 16 limits the extent within which the ordinary symbols may be used, cannot easily be decided. But it suggests limitations which may be adopted during the present discussion, and leave an ample field for curious investigation.

The distributive principle of multiplication may be adopted; namely, the principle that the product of an algebraic sum of factors into or by a common factor, is equal to the corresponding algebraic sum of the individual products of the various factors into or by the common factor; and it is expressed by the equations
$$(A \pm B)C = AC \pm BC.$$
$$C(A \pm B) = CA \pm CB.$$

32. *The associative principle of multiplication* may be adopted; namely, that the product of successive multiplications is not affected by the order in which the multiplications are performed, provided there is no change in the relative position of the factors; and it is expressed by the equations
$$ABC = (AB)C = A(BC).$$
This is quite an important limitation, and the algebras which are subject to it will be called *associative*.

33. The principle that the value of a product is not affected by the relative position of the factors is called *the commutative principle*, and is expressed by the equation

$$AB = BA.$$

This principle is *not* adopted in the present investigation.

34. An algebra in which every expression is reducible to the form of an algebraic sum of terms, each of which consists of a single *letter* with a quantitative coefficient, is called *a linear algebra*.* Such are all the algebras of the present investigation.

35. Wherever there is a limited number of independent conceptions, a linear algebra may be adopted. For a combination which was not reducible to such an algebraic sum as those of linear algebra, would be to that extent independent of the original conceptions, and would be an independent conception additional to those which were assumed to constitute the elements of the algebra.

36. An algebra in which there can be complete interchange of its independent units, without changing the formulae of combination, is a *completely symmetrical algebra;* and one in which there may be a partial interchange of its units is *partially symmetrical*. But the term symmetrical should not be applied, unless the interchange is more extensive than that involved in the distributive and commutative principles. An algebra in which the interchange is effected in a certain order which returns into itself is a *cyclic algebra*.

Thus, quaternions is a cyclic algebra, because in any of its fundamental equations, such as

$$i^2 = -1$$
$$ij = -ji = k$$
$$ijk = -1$$

there can be an interchange of the letters in the order i, j, k, i, each letter being changed into that which follows it. The double algebra in which

$$i^2 = i, \quad ij = i$$
$$j^2 = j, \quad ji = j$$

---

* In the various algebras of De Morgan's "Triple Algebra," the distributive, associative and commutative principles were all adopted, and they were all linear. [De Morgan's algebras are "semi-algebras." See Cambridge Phil. Trans., viii, 241.] [C. S. P.]

is cyclic because the letters are interchangeable in the order i, j, i. But neither of these algebras is commutative.

37. When an algebra can be reduced to a form in which all the letters are expressed as powers of some one of them, it may be called a *potential algebra*. If the powers are all squares, it may be called *quadratic*; if they are cubes, it may be called *cubic*; and similarly in other cases.

Linear Associative Algebra.

38. *All the expressions of an algebra are distributive, whenever the distributive principle extends to all the letters of the alphabet.*

For it is obvious that in the equation

$$(i+j)(k+l) = ik + jk + il + jl$$

each letter can be multiplied by an integer, which gives the form

$$(ai + bj)(ck + dl) = acik + bcjk + adil + bdjl,$$

in which a, b, c and d are integers. The integers can have the ratios of any four real numbers, so that by simple division they can be reduced to such real numbers. Other similar equations can also be formed by writing for a and b, a_1 and b_1, or for c and d, c_1 and d_1, or by making both these substitutions simultaneously. If then the two first of these new equations are multiplied by J and the last by -1; the sum of the four equations will be the same as that which would be obtained by substituting for a, b, c and d, $a + Ja_1$, $b + Jb_1$, $c + Jc_1$ and $d + Jd_1$. Hence a, b, c and d may be any numbers, real or imaginary, and in general whatever mixtures A, B, C and D may represent of the original units under the form of an algebraic sum of the letters i, j, k, &c., we shall have

$$(A + B)(C + D) = AC + BC + AD + BD,$$

which is the complete expression of the distributive principle.

39. *An algebra is associative whenever the associative principle extends to all the letters of its alphabet.*

For if
$$A = \Sigma(ai) = ai + a_1 j + a_2 k + \&c.$$
$$B = \Sigma(bi) = bi + b_1 j + b_2 k + \&c.$$
$$C = \Sigma(ci) = ci + c_1 j + c_2 k + \&c.$$

it is obvious that
$$AB = \Sigma(ab_1ij)$$
$$BC = \Sigma(bc_1ij)$$
$$(AB)C = \Sigma(ab_1c_1ijk) = A(BC) = ABC$$
which is the general expression of the associative principle.

40. *In every linear associative algebra, there is at least one idempotent or one nilpotent expression.*

Take any combination of letters at will and denote it by A. Its square is generally independent of A, and its cube may also be independent of A and A^2. But the number of powers of A that are independent of A and of each other, cannot exceed the number of letters of the alphabet; so that there must be some least power of A which is dependent upon the inferior powers. The mutual dependence of the powers of A may be expressed in the form of an equation of which the first member is an algebraic sum, such as
$$\Sigma_m(a_m A^m) = 0.$$
All the terms of this equation that involve the square and higher powers of A may be combined and expressed as BA, so that B is itself an algebraic sum of powers of A, and the equation may be written
$$BA + a_1 A = (B + a_1)A = 0.$$
It is easy to deduce from this equation successively
$$(B + a_1) A^m = 0$$
$$(B + a_1) B = 0$$
$$\left(-\frac{B}{a_1}\right)^2 = -\frac{B}{a_1}$$
so that $-\dfrac{B}{a_1}$ is an idempotent expression. But if a_1 vanishes, this expression becomes infinite, and instead of it we have the equation
$$B^2 = 0$$
so that B is a nilpotent expression.

41. When there is *an idempotent expression* in a linear associative algebra, it can be assumed as one of the independent units, and be represented by *one of the letters of the alphabet;* and it may be called *the basis.*

The remaining units can be so selected as to be separable into four distinct groups.

With reference to the basis, the units of the first group are *idemfactors;* those of the second group are *idemfaciend and nilfacient;* those of the third group are *idemfacient and nilfaciend;* and those of the fourth group are *nilfactors.*

First. The possibility of the selection of all the remaining units as idemfaciend or nilfaciend is easily established. For if i is the idempotent base, its definition gives
$$i^2 = i.$$

The product by the basis of another expression such as A may be represented by B, so that
$$iA = B,$$
which gives
$$iB = i^2 A = iA = B$$
$$i(A - B) = iA - iB = B - B = 0,$$
whence it appears that B is idemfaciend and $A - B$ is nilfaciend. In other words, A is divided into two parts, of which one is idemfaciend and the other is nilfaciend; but either of these parts may be wanting, so as to leave A wholly idemfaciend or wholly nilfaciend.

Secondly. The still farther subdivision of these portions into idemfacient and nilfacient is easily shown to be possible by this same method, with the mere reversal of the relative position of the factors. Hence are obtained the required four groups.

The basis itself may be regarded as belonging to the first group.

42. Any algebraic sum of the letters of a group is an expression which belongs to the same group, and may be called *factorially homogeneous*.

43. *The product of two factorially homogeneous expressions, which does not vanish, is itself factorially homogeneous, and its faciend name is the same as that of its facient, while its facient name is the same as that of its faciend.*

Thus, if A and B are, each of them, factorially homogeneous, they satisfy the equations
$$i(AB) = (iA)B,$$
$$(AB)i = A(Bi),$$
which shows that the nature of the product as a faciend is the same as that of the facient A, and its nature as a facient is the same as that of the faciend B.

44. *Hence, no product which does not vanish can be commutative unless both its factors belong to the same group.*

45. *Every product vanishes, of which the faciend is idemfacient while the faciend is nilfaciend; or of which the facient is nilfacient while the faciend is idemfaciend.* For in either case this product involves the equation

$$AB = (Ai)B = A(iB) = 0.$$

46. The combination of the propositions of §§ 43 and 45 is expressed in the following form of a multiplication table. In this table, each factor is expressed by two letters, of which the first denotes its name as a faciend and the second as a facient. The two letters are d and n, of which d stands for *idem* and n for *nil*. The facient is written in the left hand column of the table and the faciend in the upper line. The character of the product, when it does not vanish, is denoted by the combination of letters, or when it must vanish, by the zero, which is written upon the same line with the facient and in a column under the faciend.

	dd	dn	nd	nn
dd	dd	dn	0	0
dn	0	0	dd	dn
nd	nd	nn	0	0
nn	0	0	nd	nn

47. It is apparent from the inspection of this table, that *every expression which belongs to the second or third group is nilpotent.*

48. It is apparent that *all commutative products which do not vanish are restricted to the first and fourth groups.*

49. It is apparent that every continuous product which does not vanish, has the same faciend name as its first facient, and the same facient name as its last faciend.

50. Since the products of the units of a group remain in the group, they cannot serve as the bond for uniting different groups, which are the necessary conditions of a pure algebra. Neither can the first and fourth groups be connected by direct multiplication, because the products vanish. *The first and fourth groups, therefore, require for their indissoluble union into a pure algebra that there should be units in each of the other two groups.*

51. In an algebra which has more than two independent units, it cannot happen that all the units except the base belong to the second or to the third group. For in this case, each unit taken with the base would constitute a double algebra, and there could be no bond of connection to prevent their separation into distinct algebras.

52. *The units of the fourth group are subject to independent discussion, as if they constituted an algebra of themselves.* There must be in this group an idempotent or a nilpotent unit. *If there is an idempotent unit, it can be adopted as the basis of this group, through which the group can be subdivided into subsidiary groups.*

The idempotent unit of the fourth group can even be made the basis of the whole algebra, and the first, second and third groups will respectively become the fourth, third and second groups for the new basis.

53. *When the first group comprises any units except the basis, there is besides the basis another idempotent expression, or there is a nilpotent expression.* By a process similar to that of § 40 and a similar argument, it may be shown that for any expression A, which belongs to the first group, there is some least power which can be expressed by means of the basis and the inferior powers in the form of an algebraic sum. This condition may be expressed by the equation

$$\Sigma_m (a_m A^m) + bi = 0.$$

If then h is determined by the ordinary algebraic equation

and if
$$\Sigma_m (a_m h^m) + b = 0,$$

$$A_1 = A - hi$$

is substituted for A, an equation is obtained between the powers of A, from which an idempotent expression, B, or else a nilpotent expression, can be deduced precisely as in § 40.*

54. *When there is a second idempotent unit in the first group, the basis can be changed so as to free the first group from this second idempotent unit.*

Thus if i is the basis, and if j is the second idempotent unit of the first group, the basis can be changed to

---

* The equation in h may have no algebraic solution, in which case the new idempotent or nilpotent would not be a direct algebraic function of i and A. [C. S. P.]

PEIRCE: *Linear Associative Algebra.*

$$i_1 = i - j;$$

and with this new basis, j passes from the first to the fourth group. For *First*, the new basis is idempotent, since

$$i_1^2 = (i-j)^2 = i^2 - 2ij + j^2 = i - j = i_1;$$

and *secondly*, the idempotent unit j passes into the fourth group, since

$$i_1 j = (i-j)j = ij - j^2 = j - j = 0,$$
$$j i_1 = j(i-j) = ji - j^2 = j - j = 0.$$

55. *With the preceding change of basis, expressions may pass from idemfacient to nilfacient, or from idemfaciend to nilfaciend, but not the reverse.*

For *first*, if A is nilfacient with reference to the original basis, it is also, by § 45, nilfacient with reference to the new basis; or if it is nilfaciend with reference to the original basis, it is nilfaciend with reference to the new basis.

Secondly, all expressions which are idemfacient with reference to the original basis, can, by the process of § 41, be separated into two portions with reference to the new basis, of which portions one is idemfacient and the other is nilfacient; so that the idemfacient portion remains idemfacient, and the remainder passes from being idemfacient to being nilfacient. The same process may be applied to the faciends with similar conclusions.

56. It is evident, then, that each group* can be reduced so as not to contain more than one idempotent unit, which will be its basis. In the groups which bear to the basis the relations of second and third groups, there are only nilpotent expressions.

57. *In a group or an algebra which has no idempotent expression, all the expressions are nilpotent.*

Take any expression of this group or algebra and denote it by A. If no power of A vanished, there must be, as shown in § 40, some equation between the powers of A of the form

$$\Sigma_m a_m A^m = 0,$$

in which a_1 must vanish, or else there would be an idempotent expression as is shown in § 40, which is contrary to the present hypothesis. If then m_0 denote

---

* That is, the first group as well as each of the subsidiary groups of § 52. [C. S. P.]

the exponent of the least power of A that entered into this equation, and $m_0 + h$ the exponent of the highest power that occurred in it, the whole number of terms of the equation would be, at most, $h + 1$. If, now, the equation were multiplied successively by A and by each of its powers as high as that of which the exponent is $(m_0 - 1)h$, this highest exponent would denote the number of new equations which would be thus obtained. If, moreover,

$$B = A^{m_0},$$

then the highest power of A introduced into these equations would be

$$A^{(m_0-1)h + m_0 + h} = A^{m_0(h+1)} = B^{h+1}.$$

The whole number of powers of A contained in the equations would be $m_0 h + 1$, and $h + 1$ of these would always be integral powers of B; and there would remain $(m_0 - 1)h$ in number which were not integral powers of B. There would be, therefore, equations enough to eliminate all the powers of A that were not integral powers of B and still leave an equation between the integral powers of B; and this would generally include the first power of B. From this equation, an idempotent expression could be obtained by the process of § 40, which is contrary to the hypothesis of the proposition.

Therefore it cannot be the case that there is any equation such as that here assumed; and therefore there can be no expression which is not nilpotent. The few cases of peculiar doubt can readily be solved as they occur; but they always must involve the possibility of an equation between fewer powers of B than those in the equation in A.*

58. *When an expression is nilpotent, all its powers which do not vanish are mutually independent.*

Let A be the nilpotent expression, of which the nth power is the highest which does not vanish. There cannot be any equation between these powers of the form

$$\Sigma_m a_m A^m = 0.$$

---

* In saying that the equation in B will *generally* include the first power of B, he intends to waive the question of whether this always happens. For, he reasons, if this is not the case then the equation in B is to be treated just as the equation in A has been treated, and such repetitions of the process must ultimately produce an equation from which either an idempotent expression could be found, or else A would be proved nilpotent. [C. S. P.]

For if m_0 were the exponent of the lowest power of A in this equation, the multiplication of the equation by the $(n-m_0)^{\text{th}}$ power of A reduces it to

$$a_{m_n}A^n = 0, \quad a_{m_0} = 0,$$

that is, the m_0^{th} power of A disappears from the equation, or there is no least power of A in the equation, or, more definitely, there is no such equation.

59. *In a group or an algebra which contains no idempotent expression, any expression may be selected as the basis; but one is preferable which has the greatest number of powers which do not vanish.* All the powers of the basis which do not vanish may be adopted as independent units and represented by the letters of the alphabet.

A nilpotent group or algebra may be said to be of the same order with the number of powers of its basis that do not vanish, provided the basis is selected by the preceding principle. Thus, if the squares of all its expressions vanish, it is of the *first order*; if the cubes all vanish and not all the squares, it is *of the second order*, and so on.

60. It is obvious that *in a nilpotent group whose order equals the number of letters which it contains, all the letters except the basis may be taken as the successive powers of the basis.*

61. In a nilpotent group, every expression, such as A, has some least power that is nilfacient with reference to any other expression, such as B, and which corresponds to what may be called *the facient order of B relatively to A*; and in the same way, there is some least power of A which is nilfaciend with reference to B, and which corresponds to *the faciend order of B relatively to A*. When the facient and faciend orders are treated of irrespective of any especial reference, *they must be referred to the base.*

The facient order of a product which does not vanish, is not higher than that of its facient; and the faciend order is not higher than that of its faciend.

62. After the selection of the basis of a nilpotent group, some one from among the expressions which are independent of the basis may be selected by the same method by which the basis was itself selected, *which, together with all its powers that are independent of the basis, may be adopted as new letters;* and again, from the independent expressions which remain, *new letters may be selected by the same process, and so on until the alphabet is completed.* In making these selections, regard should be had to the factorial orders of the products.

63. In every nilpotent group, *the facient order of any letter which is independent of the basis can be assumed to be as low as the number of letters which are independent of the basis.*

Thus, if the number of letters which are independent of the basis is denoted by n', and if n is the order of the group (and for the present purpose it is sufficient to regard n' as being less than n), it is evident that any expression, A, with its successive products by the powers of the basis i, as high as the n'^{th}, and the powers of the basis which do not vanish, cannot all be independent of one another; so that there must be an equation of the form

$$\sum_1^n a_m i^m + \sum_0^{n'} b_m i^m A = 0.$$

Accordingly, it is easy to see that there is always a value of A_1 of the form

$$A_1 = A - \sum_1^n h_m i^m$$

which will give

$$i^m A_1 = 0,$$

which corresponds to the condition of this section.

There is a similar condition which holds in every selection of a new letter by the method of the preceding section.

64. *In a nilpotent group, the order of which is less by unity than the number of letters, the letter which is independent of the basis and its powers may be so selected that its product into the basis shall be equal to the highest power of the basis which does not vanish, and that its square shall either vanish or shall also be equal to the highest power of the basis that does not vanish.* Thus, if the basis is i, and if the order of the algebra is n, and if j is the remaining letter, it is obvious, from § 63, that j might have been assumed such that

$$ij = 0,$$

which gives

$$iji = ij^2 = 0;$$

and therefore,

$$ji = ai^n + bj,$$
$$j^2 = a'i^n + bj,$$
$$0 = ji^{n+1} = bji^n = b^n ji = b,$$
$$ji = ai^n,$$
$$j^2 i = aji^n = 0 = b'j^2 = b',$$
$$j^2 = a'i^n;$$

so that if
$$h = \left(\frac{a'}{a^2}\right)^{\frac{1}{n-2}}, \quad k = \left(\frac{a'^{n-1}}{a^n}\right)^{\frac{1}{n-2}}$$
$$j_1 = \frac{j}{k}, \quad i_1 = \frac{i}{h},$$
we have
$$j_1 i_1 = i_1^n = j_1^2,$$

and i_1 and j_1 can be substituted for i and j, which conforms to the proposition enunciated.

It must be observed, however, that the analysis needs correction when the group is of the second order.

65. *In a nilpotent group of the first order, the sign of a product is merely reversed by changing the order of its factors.* Thus, if
$$A^2 = B^2 = (A + B)^2 = 0$$
it follows by development, that
$$(A + B)^2 = A^2 + AB + BA + B^2 = AB + BA = 0$$
$$BA = -AB,$$
which is the proposition enunciated.

66. *In general, in any nilpotent group of the nth order, if (A^s, B^t) denotes the sum of all possible products of the form*
$$A^p B^q \; A^{p'} B^{q'} \; A^{p''} B^{q''} \ldots$$
in which
$$\Sigma p = s, \quad \Sigma q = t,$$
and if
$$s + t = n + 1,$$
it will be found that
$$(A^s, B^t) = 0.$$
For since
$$(A + xB)^{n+1} = 0$$
whatever be the value of x, the multiplier of each power of x must vanish, which gives the proposed equation
$$(A^s, B^t) = 0.$$

67. *In the first group of an algebra, having an idempotent basis, all the expressions except the basis may be assumed to be nilpotent.* For, by the same argument as that of §53, any equation between an expression and its successive powers and the basis must involve an equation between another expression which is

easily defined and its successive powers without including the basis. But it follows from the argument of § 57, that such an equation indicates a corresponding idempotent expression; whereas it is here assumed that, in accordance with § 56, each group has been brought to a form which does not contain any other idempotent expression than the basis. It must be, therefore, that all the other expressions are nilpotent.

68. *No product of expressions in the first group of an algebra having an idempotent basis, contains a term which is a multiple of the basis.*

For, assume the equation
$$AB = -xi + C,$$
in which A, B and C are nilpotents of the orders m, n and p, respectively. Then,
$$0 = A^{m+1}B = -xA^m + A^m C$$
$$A^m C = xA^m$$
$$0 = A^m C^{p+1} = xA^m C^p = x^2 A^m C^{p-1} = x^{p+1} A^m = x,$$
that is, the term $-xi$ vanishes from the product AB.

69. It follows, from the preceding section, that *if the idempotent basis were taken away from the first group of which it is the basis, the remaining letters of the first group would constitute by themselves a nilpotent algebra.*

Conversely, *any nilpotent algebra may be converted into an algebra with an idempotent basis, by the simple annexation of a letter idemfaciend and idemfacient with reference to every other.**

70. However incapable of interpretation the nilfactorial and nilpotent expressions may appear, they are obviously an essential element of the calculus of linear algebras. Unwillingness to accept them has retarded the progress of discovery and the investigation of quantitative algebras. But the idempotent basis seems to be equally essential to actual interpretation. The purely nilpotent algebra may therefore be regarded as an ideal abstraction, which requires the introduction of an idempotent basis, to give it any position in the real universe. In the subsequent investigations, therefore, the purely nilpotent algebras must be regarded as the first steps towards the discovery of algebras of a higher degree resting upon an idempotent basis.

---

* That every such algebra must be a pure one is plain, because the algebra (a_2) is so. [C. S. P.]

71. Sufficient preparation is now made for the

INVESTIGATION OF SPECIAL ALGEBRAS.

The following notation will be adopted in these researches. Conformably with §9, the letters of the alphabet will be denoted by i, j, k, l, m and n. To these letters will also be respectively assigned the numbers 1, 2, 3, 4, 5 and 6. Moreover, their coefficients in an algebraic sum will be denoted by the letters a, b, c, d, e and f. Thus, the product of any two letters will be expressed by an algebraic sum, and below each coefficient will be written in order the numbers which are appropriate to the factors. Thus,

while
$$jl = a_{24} i + b_{24} j + c_{24} k + d_{24} l + e_{24} m + f_{24} n,$$
$$lj = a_{42} i + b_{42} j + c_{42} k + d_{42} l + e_{42} m + f_{42} n.$$

In the case of a square, only one number need be written below the coefficient, thus
$$l^2 = a_3 i + b_3 j + c_3 k + d_3 l + e_3 m + f_3 n.$$

The investigation simply consists in the determination of the values of the coefficients, corresponding to every variety of linear algebra; and the resulting products can be arranged in a tabular form which may be called the multiplication-table of the algebra. Upon this table rests all the peculiarity of the calculus. In each of the algebras, it admits of many transformations, and much corresponding speculation. The basis will be denoted by i.

72. The distinguishing of the successive cases by the introduction of numbers will explain itself, and is an indispensable protection from omission of important steps in the discussion.

SINGLE ALGEBRA.

Since in a single algebra there is only one independent unit, it requires no distinguishing letter. It is also obvious that there can be no single algebra which is not associative and commutative. Single algebra has, however, two cases:

[1], when its unit is idempotent;
[2], when it is nilpotent.

[1]. The defining equation of this case is
$$i^2 = i.$$

This algebra may be called (a_1) and its multiplication table is *

(a_1)	i
i	i

[2]. The defining equation of this case is

$$i^2 = 0.$$

This algebra may be called (b_1) and its multiplication table is †

(b_1)	i
i	0

DOUBLE ALGEBRA.

There are two cases of double algebra:

[1], when it has an idempotent expression;
[2], when it is nilpotent.

[1]. The defining equation of this case is

$$i^2 = i.$$

By §§ 41 and 50, there are two cases:

[1²], when the other unit belongs to the first group;
[12], when it is of the second group.

The hypothesis that the other unit belongs to the third group is a virtual repetition of [12].

[1²]. The defining equations of this case are

$$ij = ji = j.$$

It follows from §§ 67 and 69, that there is a double algebra derived from (b_1) which may be called (a_2), of which the multiplication table is ‡

---

* This algebra may be represented by $i = A : A$ in the logic of relatives. See Addenda. [C. S. P.]
† This algebra takes the form $i = A : B$, in the logic of relatives. [C. S. P.]
‡ This algebra may be put in the form $i = A : A + B : B$, $j = A : B$. [C. S. P.]

(a_2)	i	j
i	i	j
j	j	0

[12]. The defining equations of this case are, by § 41,
$$ij = j,\ ji = 0;$$
whence, by § 46,
$$j^2 = 0.$$

A double algebra is thus formed, which may be called (b_2), of which the multiplication table is*

(b_2)	i	j
i	i	j
j	0	0

[2]. The defining equation of this case is
$$i^n = 0,$$
in which n is the least power of i which vanishes. There are two cases:

$$[21],\ \text{when}\ n = 3;$$
$$[2^2],\ \text{when}\ n = 2.$$

[21]. The defining equation of this case is
$$i^3 = 0,$$
and by § 60,
$$i^2 = j.$$

This gives a double algebra which may be called (c_2), its multiplication table being †

---

*This algebra may be put in the form $i = A:A, j = A:B$. [C. S. P.]
† In relative form, $i = A:B + B:C, j = A:C$. [C. S. P.]

(c_2)	i	j
i	j	0
j	0	0

[2^2]. The defining equations of this case are

$$i^2 = j^2 = 0,$$

and it follows from §§ 64 and 65 that

$$ij = ji = 0,$$

so that there is no pure algebra in this case.*

TRIPLE ALGEBRA.

There are two cases:

[1], when there is an idempotent basis;
[2], when the basis is nilpotent.

[1]. The defining equation of this case is

$$i^2 = i.$$

There are, by §§ 41, 50 and 51, three cases:

[1^2], when j and k are both in the first group;
[12], when j is in the first, and k in the second group;
[13], when j is in the second, and k in the third group.

The case of j being in the first, and k in the third group, is a virtual repetition of [12].

[1^2]. The defining equations of this case are

$$ij = ji = j, \quad ik = ki = k.$$

---

*This case takes the form $i = A : B$, $j = C : D$. [C. S. P.]

It follows from §§ 67 and 69, that the only algebra of this case may be derived from (c_2); it may be called (a_3), and its multiplication table is *

(a_3)	i	j	k
i	i	j	k
j	j	k	0
k	k	0	0

[12]. The defining equations of this case are

$$ji = ij = j, \quad ik = k, \quad ki = 0;$$

whence, by §§ 46 and 67,

$$j^2 = k^2 = kj = 0, \quad jk = c_{23}k,$$
$$j^2k = 0 = c_{23}jk = c_{23}^2 k = c_{23} = jk,$$

and there is no pure algebra in this case.†

[13]. The defining equations of this case are

$$ij = j, \quad ki = k, \quad ji = ik = 0;$$

whence, by § 46,

$$j^2 = k^2 = kj = 0, \quad jk = a_{23}i,$$
$$jkj = 0 = a_{23}j = a_{23} = jk,$$

and there is no pure algebra in this case.‡

[2]. The defining equation of this case is

$$i^n = 0,$$

in which n is the lowest power of i that vanishes.

There are three cases:

[21], when $n = 4$;
[2²], when $n = 3$;
[23], when $n = 2$.

---

*In relative form, $i = A : A + B : B + C : C$, $j = A : B + B : C$, $k = A : C$. [C. S. P.]

† That is to say, i and j by themselves form the algebra a_2, and i and k by themselves constitute the algebra b_2, while the products of j and k vanish. Thus, the three letters are not indissolubly bound together into one algebra. In relative form, this case is, $i = A : A + B : B$, $j = A : B$, $k = A : C$. [C. S. P.]

‡ In relative form, $i = A : A + D : D$, $j = A : B$, $k = C : D$. [C. S. P.]

[21]. The defining equation of this case is
$$i^4 = 0,$$
and by § 60
$$i^2 = j, \quad i^3 = k.$$

This gives a triple algebra which may be called (b_3), the multiplication table being *

(b_3)	i	j	k
i	j	k	0
j	k	0	0
k	0	0	0

[2²]. The defining equation of this case is
$$i^3 = 0,$$
and by §§ 59 and 64, observing the exception,
$$i^2 = j, \quad ik = 0,$$
$$ki = b_{31}j, \quad k^2 = b_3 j.$$

There is no pure algebra when b_{31} vanishes,† and there are two cases:

[2²1], when b_3 does not vanish;
[2³], when b_3 vanishes.

[2²1]. The defining equation of this case can, without loss of generality, be reduced to
$$k^2 = j.$$

This gives a triple algebra which may be called (c_3), the multiplication table being ‡

---

* In relative form, $i = A:B + B:C + C:D$, $j = A:C + B:D$, $k = A:D$. [C. S. P.]
† This case takes the relative form, $i = A:B + B:C$, $j = A:C$, $k = b_3 A:D + D:C$. [C. S. P.]
‡ In relative form, $i = A:B + B:C$, $j = A:C$, $k = a.A:B + A:D + D:C$. [C. S. P.]

(c_3)	i	j	k
i	j	0	0
j	0	0	0
k	aj	0	j

An interesting special example of this case is afforded by $a = -2$, when

$$i(k+i) = -j$$
$$(k+i)i = j$$
$$(k+i)^2 = 0,$$

so that $k+i$ might be substituted for k, and in this form, the multiplication table of this algebra, which may be called (c'_3), is*

(c'_3)	i	j	k
i	j	0	j
j	0	0	0
k	$-j$	0	0

---

* In relative form, $i = A:B + B:C$, $j = A:C$, $k = -A:B + B:C + A:D + D:C$.

When $a = +2$, the algebra equally takes the form (c'_3), on substituting $k-i$ for k. On the other hand, provided a is neither 2 nor -2, the algebra may be put in the form

(c''_3)	i	j	k
i	0	0	$-bj$
j	0	0	0
k	$b^{-1}j$	0	0

To effect the transformation, we write $a = -b - \frac{1}{b}$ and substitute $i + bk$ and $i + \frac{1}{b}k$ for i and k, and $\left(b - \frac{1}{b}\right)j$ for j. Thus the algebra (c_3) has two distinct and intransmutable species, (c'_3) and (c''_3). [C. S. P.]

[2^3]. The defining equation of this case is

$$k^2 = 0,$$

and b_{31} may be reduced to unity without loss of generality, giving a triple algebra which may be called (d_3), the multiplication table being

(d_3)	i	j	k
i	j	0	0
j	0	0	0
k	j	0	0

In this case

$$(i-k)k = 0$$
$$k(i-k) = 0$$
$$(i-k)^2 = 0,$$

so that $i-k$ may be substituted for i, and in this form the multiplication table is *

(d_3)	i	j	k
i	0	0	0
j	0	0	0
k	j	0	0

[23]. The defining equations of this case are

$$i^2 = j^2 = k^2 = 0,$$

and by the principles of §§ 63 and 65, it may be assumed that

$$ij = -ji = -ik = ki = 0,$$
$$jk = -kj = i.$$

---

* In relative form, $i = B:C$, $j = A:C$, $k = A:B$. This is the algebra of alio-relations in its typical form. [C. S. P.]

We thus get a triple algebra which may be called (c_3), its multiplication table being*

(c_3)	i	j	k
i	0	0	0
j	0	0	i
k	0	$-i$	0

QUADRUPLE ALGEBRA.

There are two cases:

[1], when there is an idempotent basis;
[2], when the base is nilpotent.

[1]. The defining equation of this case is

$$i^2 = i.$$

There are six cases:

[1^2], when j, k, and l, are all in the first group;
[12], when j and k are in the first, and l in the second group;
[13], when j is in the first, and k and l in the second group;
[14], when j is in the first, k in the second, and l in the third group;
[15], when j and k are in the second, and l in the third group;
[16], when j is in the second, k in the third, and l in the fourth group.

The other cases are excluded by §§ 50 and 51, or are obviously virtual repetitions of those which are given.

[1^2]. The defining equations of this case are

$$ij = ji = j, \quad ik = ki = k, \quad il = li = l,$$

and from §§ 60 and 69, the algebras (b_3), (c_3), (d_3), and (e_3), give quadruple algebras which may be named respectively (a_4), (b_4), (c_4), and (d_4), their multiplication tables being

---
* In relative form, $i = A:D$, $j = A:B - C:D$, $k = A:C + B:D$. This is the algebra of alternate numbers. [C. S. P.]

(a_4)

	i	j	k	l
i	i	j	k	l
j	j	k	l	0
k	k	l	0	0
l	l	0	0	0

(b_4)

	i	j	k	l
i	i	j	k	l
j	j	k	0	0
k	k	0	0	0
l	l	ak	0	k

(c_4)

	i	j	k	l
i	i	j	k	l
j	j	k	0	0
k	k	0	0	0
l	l	k	0	0

(d_4)

	i	j	k	l
i	i	j	k	l
j	j	0	0	0
k	k	0	0	j
l	l	0	$-j$	0

The special case (c'_3) gives a corresponding special case of (b_4), which may be called (b'_4), of which the multiplication table is

(b'_4)

	i	j	k	l
i	i	j	k	l
j	j	k	0	k
k	k	0	0	0
l	l	$-k$	0	0

The second form of (d_3) gives a corresponding second form of (c_4), of which the multiplication table is

(c_4)	i	j	k	l
i	i	j	k	l
j	j	0	0	0
k	k	0	0	0
l	l	k	0	0

[12]. The defining equations of this case are

$$ij = ji = j, \quad ik = ki = k, \quad il = l, \quad li = 0,$$

and it follows from §§ 67 and 69, that (c_2) gives

$$j^2 = k, \quad jk = kj = k^2 = 0,$$

and from § 46,

$$lj = lk = l^2 = 0, \quad jl = d_{24}l, \quad kl = d_{34}l;$$

whence

$$j^2l = kl = d_{24}jl = d_{24}^2 l,$$
$$jkl = d_{24}^2 jl = d_{24}^3 l = 0 = d_{24} = jl = kl,$$

and there is no pure algebra in this case.*

[13]. The defining equations of this case are

$$ij = ji = j, \quad ik = k, \quad il = l, \quad ki = li = 0,$$

which give by §§ 46 and 67

$$0 = j^2 = k^2 = kl = lk = l^2 = kj = lj;$$

and it may be assumed that

$$jk = l, \text{ whence } jl = 0.$$

This gives a quadruple algebra which may be called (e_4), its multiplication table being †

---

*In relative form, $i = A:A + B:B + C:C + D:D$, $j = A:B + B:C$, $k = A:C$, $l = D:C$. [C. S. P.]

† In relative form, $i = A:A + B:B$, $j = A:B$, $k = B:C$, $l = A:C$. [C. S. P.]

(c_4)	i	j	k	l
i	i	j	k	l
j	j	0	l	0
k	0	0	0	0
l	0	0	0	0

[14]. The defining equations of this case are

$$ij = ji = j, \quad ik = k, \quad li = l, \quad ki = il = 0;$$

which give, by §§ 46 and 67,

$$0 = j^2 = jl = kj = k^2 = lk = l^2,$$
$$jk = c_{23}k, \quad lj = d_{42}l, \quad kl = a_{34}i + b_{34}j,$$
$$0 = j^2 k = c_{23} jk = c_{23}^2 k = c_{23} = jk,$$
$$0 = lj^2 = d_{42} lj = d_{42}^2 l = d_{42} = lj,$$
$$0 = jkl = a_{34} j = a_{34},$$

and b_{34} cannot be permitted to vanish,* so that it does not lessen the generality to assume

$$kl = j.$$

This gives a quadruple algebra which may be called (f_4), its multiplication table being †

(f_4)	i	j	k	l
i	i	j	k	0
j	j	0	0	0
k	0	0	0	j
l	l	0	0	0

---

* For then the algebra would split up into three double algebras. [C. S. P.]
† In relative form, $i = A:A + B:B, j = A:B, k = A:C, l = C:B$. [C. S. P.]

[15]. The defining equations of this case are
$$ij = j, \quad ik = k, \quad li = l, \quad ji = ki = il = 0,$$
which give, by § 46,
$$0 = j^2 jk = kj = k^2 = lj = lk = l^2,$$
$$jl = a_{24} i, \quad kl = a_{34} i,$$
$$0 = jlj = a_{24} j = a_{24} = jl,$$
$$0 = klk = a_{34} k = a_{34} = kl,$$
and there is no pure algebra in this case.*

[16]. The defining equations of this case are
$$ij = j, \quad ki = k, \quad ji = ik = il = li = 0,$$
which give, by § 46,

$0 = j^2 = k^2 = kl = lj,$
$jk = a_{23} i, \quad jl = b_{24} j, \quad kj = d_{32} l, \quad lk = c_{43} k, \quad l^2 = d_4 l,$
$jkj = a_{23} j = d_{32} b_{24} j, \quad jlk = a_{23} b_{24} i = a_{23} c_{43} i, \quad jl^2 = b_{24}^2 j = b_{24} d_4 j,$
$kjk = a_{23} k = c_{43} d_{32} k, \quad kjl = d_{32} d_4 l = b_{24} d_{32} l,$
$lkj = d_{32} d_4 l = c_{43} d_{32} l, \quad l^2 k = c_{43}^2 k = c_{43} d_4 k, \quad a_{23} = b_{24} d_{32},$
$0 = a_{23}(c_{43} - b_{24}) = b_{24}(b_{24} - d_4) = d_{32}(b_{24} - d_4) = d_{32}(c_{43} - d_4) = c_{43}(c_{43} - d_4).$

There are two cases:

[161], when d_{32} does not vanish;
[162], when d_{32} vanishes.

[161]. The defining equation of this case can be reduced to
$$d_{32} = 1,$$
which gives
$$a_{23} = b_{24} = c_{43} = d_4.$$
There are two cases:

[161²], when d_4 does not vanish;
[1612], when d_4 vanishes.

[161²]. The defining equation of this case can be reduced to
$$d_4 = 1,$$
which gives
$$jk = i, \quad jl = j, \quad lk = k, \quad l^2 = l;$$

---

*In relative form, $i = A : A, j = A : B, k = A : C, l = D : A$. There are three double algebras of the form (b_4). [C. S. P.]

and there is a quadruple algebra which may be called (g_4), its multiplication table being

(g_4)	i	j	k	l
i	i	j	0	0
j	0	0	i	j
k	k	l	0	0
l	0	0	k	l

This is a form of *quaternions*.*

[1612]. The defining equation of this case is

$$d_4 = 0,$$

which gives

$$jk = jl = lk = l^2 = 0,$$

---

*In relative form, $i = A:A$, $j = A:B$, $k = B:A$, $l = B:B$. This algebra exhibits the general system of relationship of individual relatives, as is shown in my paper in the ninth volume of the Memoirs of the American Academy of Arts and Sciences. In a space of four dimensions, a vector may be determined by means of its rectangular projections on two planes such that every line in the one is perpendicular to every line in the other. Call these planes the A-plane and the B-plane, and let v be any vector. Then, iv is the projection of v upon the A-plane, and lv is its projection upon the B-plane. Let each direction in the A-plane be considered as to correspond to a direction in the B-plane in such a way that the angle between two directions in the A-plane is equal to the angle between the corresponding directions in the B-plane. Then, jv is that vector in the A-plane which corresponds to the projection of v upon the B-plane, and kv is that vector in the B-plane which corresponds to the projection of v upon the A-plane.

Professor Peirce showed that we may take i_1, j_1, k_1, as three such mutually perpendicular vectors in ordinary space, that $i = \frac{1}{2}(1 - ji_1)$, $j = \frac{1}{2}(i_1 - jk_1)$, $k = \frac{1}{2}(-j_1 - jk_1)$, $l = \frac{1}{2}(1 + ji)$. [See, also, Spottiswoode, Proceedings of the London Mathematical Society, iv, 156. Cayley; in his Memoir on the Theory of Matrices (1858), had shown how a quaternion may be represented by a dual matrix.] Thus i, j, k, l, have all zero tensors, and j and k are vectors. In the general expression of the algebra, $q = xi + yj + zk + wl$, if $x + w = 1$ and $yz = x - x^2$, we have $q^2 = q$; if $x = -w = \sqrt{-yz}$, then $q^2 = 0$. The expression $i + l$ represents scalar unity, since it is the universal idemfactor. We have, also, $Sq = \frac{1}{2}(x+w)(i+l)$, $Vq = \frac{1}{2}(x-w)i + yj + zk + \frac{1}{2}(w-x)l$, $Tq = \sqrt{xw - yz}(i+l)$.

The resemblance of the multiplication table of this algebra to the symbolical table of §46 merits attention. [C. S. P.]

and there is a quadruple algebra which may be called (h_4), its multiplication table being*

(h_4)	i	j	k	l
i	i	j	0	0
j	0	0	0	0
k	k	l	0	0
l	0	0	0	0

[162]. The defining equation of this case is

$$d_{32} = 0,$$

which gives

$$a_{23} = 0,$$

and there can be no pure algebra for it.†

[2]. The defining equation of this case is

$$i^n = 0.$$

There are four cases: [2^1], when $n = 5$;
[2^2], when $n = 4$;
[2^3], when $n = 3$;
[2^4], when $n = 2$.

[2^1]. The defining equation of this case is

$$i^5 = 0,$$

and by § 60, $\quad i^2 = j, \quad i^3 = k, \quad i^4 = l.$

This gives a quadruple algebra which may be called (i_4), its multiplication table being ‡

---

* In relative form, $i = A:A$, $j = A:B$, $k = C:A$, $l = C:B$. [C. S. P.]
† In this case, $i = A:A$, $l = d_4(B:B + C:C)$, $j = A:B$ or $= A:D$, $k = C:A$ or $= E:A$. [C. S. P.]
‡ In relative form, $i = A:B + B:C + C:D + D:E$, $j = A:C + B:D + C:E$, $k = A:D + B:E$, $l = A:E$. [C. S. P.]

(i_4)	i	j	k	l
i	j	k	l	0
j	k	l	0	0
k	l	0	0	0
l	0	0	0	0

[2^a]. The defining equation of this case is
$$i^4 = 0,$$
and by § 59,
$$i^2 = j, \quad i^3 = k.$$
There are then, by § 64, two quadruple algebras, which may be called (j_4) and (k_4), their multiplication tables being*

(j_4)	i	j	k	l
i	j	k	0	0
j	k	0	0	0
k	0	0	0	0
l	k	0	0	k

and

(k_4)	i	j	k	l
i	j	k	0	0
j	k	0	0	0
k	0	0	0	0
l	k	0	0	0

[23]. The defining equation of this case is
$$i^3 = 0,$$
and by § 59
$$i^2 = j,$$
and it may be assumed from the principle of § 63 that
$$ik = 0,$$
which gives
$$jk = 0.$$

---

*In either of these algebras, $i = A:B + B:C + C:D$, $j = A:C + B:D$, $k = A:D$; and in (j_4) $l = A:E + E:D + A_1:C$, while in (k_4) $l = A:C$. [C. S. P.]

There are two cases: [231], when $il = k$;
  [232], when $il = 0$.

[231]. The defining equation of this case is

$$il = k,$$

which gives

$jl = i^2 l = ik = 0,$
$ki = a_{31}i + b_{31}j + c_{31}k + d_{31}l,$
$0 = iki = a_{31}j + d_{31}k,$ $a_{31} = 0,$ $d_{31} = 0,$ $ki = b_{31}j + c_{31}k.$

So, because $ik^2 = 0,$ $k^2 = b_3 j + c_3 k,$

and because $ikl = 0,$ $kl = b_{34}j + c_{34}k,$ $kj = kii = c_{31}ki = b_{31}c_{31}j + c_{31}^2 k.$
$0 = kji = c_{31}^2 ki,$ $c_{31} = 0 = kj,$
$ili = ki = b_{31}j,$ $li = b_{31}i + b_{41}j + c_{41}k,$ $lj = li^2 = (b_{31} + b_{31}c_{41})j,$
$0 = k^3 = c_3 k^2 = c_3,$ $ilk = k^2 = b_3 j,$ $lk = b_3 i + b_{43}j + c_{43}k,$
$l^2 = a_4 i + b_4 j + c_4 k + d_4 l,$ $0 = l^3 = a_4 k + c_4 kl + d_4 l^2 = a_4 li + b_4 lj + c_4 lk + d_4 l^2.$

But kl contains no term in l, so that $d_4 = 0.$
$kl = il^2 = a_4 j,$ $b_{34} = a_4,$ $c_{34} = 0,$
$0 = l^3 = b_{34}k + c_4 b_{34}j,$ $b_{34} = a_4 = 0 = kl,$ $l^2 = b_4 j + c_4 k,$
$kil = k^2 = b_{31}jl = 0,$ $0 = kli = b_{31}ki = b_{31}^2 j = b_{31} = ki = lj,$
$li = b_{41}j + c_{41}k,$ $lk = lil = 0.$

There are two cases:
  [231²], when c_{41} does not vanish;
  [2312], when c_{41} vanishes.

[231²]. The defining formula of this case is

$$c_{41} \neq 0,$$

and if p is determined by the equation

$$c_{41}p^2 + (c_4 - b_{41})p = b_4,$$

we have

$$i(l + pi) = k + pj,$$
$$(l + pi)^2 = (c_4 + pc_{41})(k + pj),$$

so that $l + pi$ and $k + pj$ may be substituted respectively for l and k, which is the same as to make

$$b_4 = 0,$$

and there are two cases:

[231³], when c_4^* does not vanish;
[231²2], when c_4 vanishes.

[221³]. The defining equation of this case can be reduced to
$$l^2 = k.$$
This gives a quadruple algebra which may be called (l_4), its multiplication table being †

(l_4)	i	j	k	l
i	j	0	0	k
j	0	0	0	0
k	0	0	0	0
l	$bj+ck$	0	0	k

[231²2]. The defining equation of this case is
$$l^2 = 0.$$

---

* I. e. the new c_4, or what has been written $c_4 + pc_{41}$. In all cases, when new letters of the alphabet of the algebra are substituted, the coefficients change with them. [C. S. P.]

†When $b = 0$, $c = 1$. we have $l(i-l) = (i-l)l = 0$; so that by the substitution of $i-l$ for i, the algebra is broken up into two of the form (c_2). When $b = 0$, $c \neq 1$, on substituting $i_1 = i - l$, $j_1 = j - ck$, $k_1 = (c-1)^2 k$, $l_1 = (c-1)l$, we have $i_1^2 = j_1$, $i_1 l_1 = 0$, $l_1 i_1 = l_1^2 = k_1$; so that the algebra reduces to (r_4). When $b = 1$, $c = 0$, on putting $i_1 = i - l$, $j_1 = j - k$, we have $i_1^2 = i_1 l = 0$, $k_1 = j_1$, $l^2 = k$; so that the algebra reduces to (q_4). When $b = 1$, $c \neq 0$, on putting $i_1 = \sqrt{c^{-1}} (i-l)$, $j_1 = j + (c-1)k$, we have $i_1^2 = l^2 = k$, $i_1 l = 0$, $k_1 = j_1$; so that the algebra reduces to (p_4). When $b(b-1)(bc+b-1) \neq 0$, on putting $i_1 = (1-b)bi - (1-b)l$, $j_1 = (1-b)^2(1-b-bc)k$, $k_1 = b^2(1-b)(1-b-bc)j - b(1-b)(1-b-c+c^2b)k$, $l_1 = b(1-b)i - bcl$, we get the multiplication table of (o_4). When $b(b-1) \neq 0$, $bc+b = 1$; on putting $i_1 = b(i-l)$, $j_1 = b^2(1-b)j - b^2ck$, $k_1 = b(1-b-c)k$, $l_1 = bi - l$, we get the following multiplication table, which may replace that in the text:

(l_4)	i	j	k	l
i	j	0	0	j
j	0	0	0	0
k	0	0	0	0
l	k	0	0	0

In relative form, $i = A:B + B:C + A:D$, $j = A:C$, $k = A:E$, $e = A:B + D:E$. [C. S. P.]

There are two cases:

$[231^2 21]$, when b_{41} does not vanish;
$[231^2 2^2]$, when b_{41} vanishes.

$[231^2 21]$. The defining formula of this case is

$$b_{41} \neq 0.$$

There are two cases:

$[231^2 21^2]$, when $c_{41} + 1$ does not vanish;
$[231^2 212]$, when $c_{41} + 1$ vanishes.

$[231^2 21^2]$. The defining formula of this case is

$$c_{41} + 1 \neq 0$$

so that

$$l \frac{b_{41}i + c_{41}l}{c_{41}+1} = \frac{b_{41}^2 j + b_{41}c_{41}k}{c_{41}+1},$$

$$\frac{b_{41}i + c_{41}l}{c_{41}+1} l = \frac{b_{41}k}{c_{41}+1}, \quad \left(\frac{b_{41}i + c_{41}l}{c_{41}+1}\right)^2 = \frac{b_{41} + c_{41}}{c_{41}+1} \cdot \frac{b_{41}j + c_{41}k}{c_{41}+1}$$

so that the substitution of $\frac{b_{41}i + c_{41}l}{c_{41}+1}$, $\frac{b_{41}^2 j + b_{41}c_{41}k}{c_{41}+1}$, and $\frac{b_{41}k}{c_{41}+1}$, respectively, for i, j, and k, is the same as to assume

$$c_{41} = 0, \quad b_{41} = j,$$

which reduces this case to [2312].

$[231^2 212]$. The defining equation of this is easily reduced to

$$li = j - k.$$

This gives a quadruple algebra which may be called (m_4), its multiplication table being

(m_4)	i	j	k	l
i	j	0	0	k
j	0	0	0	0
k	0	0	0	0
l	$j-k$	0	0	0

The substitution of $i-l$ and $j-k$, respectively, for i and j transforms this algebra into one of which the multiplication table is *

(m_4)	i	j	k	l
i	0	0	0	k
j	0	0	0	0
k	0	0	0	0
l	j	0	0	0

[$231^2 2^2$]. The defining equation of this case is

$$li = c_{41}k.$$

This gives a quadruple algebra which may be called (n_4), its multiplication table being †

(n_4)	i	j	k	l
i	j	0	0	k
j	0	0	0	0
k	0	0	0	0
l	ck	0	0	0

[2312]. The defining equation of this case is

$$li = b_{41}j,$$

which gives

$$(l - b_{41}i)i = 0,$$

so that the substitution of $l - b_{41}i$ for l passes this case virtually into [232].

---

* $i = A:B+C:D, j = B:D, k = A:C, l = B:C$. [C. S. P.]
† In relative form, $i = A:B+B:C+D:E, j = A:C, k = A:E, l = B:E+cA:D$. When $c=0$ the algebra reduces to (q_4). [C. S. P.]

[232]. The defining equation of this case is
$$il = 0,$$
and it may be assumed that
$$ki = 0,$$
$$0 = jl = kj = ik^2 = k^2i = ikl = ili = ilk = lki = il^2$$
$$k^2 = b_3j + c_3k + d_3l, \quad li = b_{41}j + c_{41}k + d_{41}l,$$
$$lj = d_{41}li, \quad 0 = lji = d_{41}lj = d_{41}^2li = d_{41} = lj.$$

There are two cases:

[2321], when c_{41} does not vanish;

[232²], when c_{41} vanishes.

[2321]. The defining equation of this case is easily reduced to
$$li = k,$$
which gives
$$0 = lik = k^2 = lil = kl$$
$$lk = l^2i = a_4j + d_4k,$$
$$0 = l^3k = d_4l^2k = d_4^2lk = d_4, \quad lk = a_4j = l^2i,$$
$$l^2 = a_4i + b_4j + c_4k,$$
$$0 = l^3 = a_4k + c_4a_4j = a_4 = lk.$$

There are two cases:

[2321²], when c_4 does not vanish;

[23212], when c_4 vanishes.

[2321²]. The defining equation of this case can be reduced to
$$c_4 = 1$$
which gives a quadruple algebra which may be called (o_4), its multiplication table being*

(o_4)	i	j	k	l
i	j	0	0	0
j	0	0	0	0
k	0	0	0	0
l	k	0	0	$bj+k$

---

*In relative form, $i = A:E + E:D + B:C$, $j = A:D$, $k = A:C$, $l = A:B + B:C + bB:D$. When $b = 0$, this algebra reduces to (r_4). When $b = -1$, the substitution of $i-l$ for l reduces it to (l_4). [C. S. P.]

[23212]. The defining equation of this case is
$$l^2 = b_4 j.$$
There are two cases:
    [232121], when b_4 does not vanish;
    [23212²], when b_4 vanishes.

[232121]. The defining equation of this case can be reduced to
$$l^2 = j.$$
This gives a quadruple algebra which may be called (p_4), its multiplication table being*

(p_4)	i	j	k	l
i	j	0	0	0
j	0	0	0	0
k	0	0	0	0
l	k	0	0	j

[23212²]. The defining equation of this case is
$$l^2 = 0.$$
This gives a quadruple algebra which may be called (q_4), its multiplication table being †

(q_4)	i	j	k	l
i	j	0	0	0
j	0	0	0	0
k	0	0	0	0
l	k	0	0	0

---

* In relative form, $i = A:B + B:D + C:E$, $j = A:D$, $k = A:E$, $l = A:C + C:D$. [C. S. P.]
† In relative form, $i = A:C + C:D$, $j = A:D$, $k = B:D$, $l = B:C$. [C. S. P.]

[232^2]. The defining equation of this case is
$$li = b_{41}j$$
and we have
$$k^2 = b_3 j + c_3 k + d_3 l$$
$$kl = b_{34} j + c_{34} k + d_{34} l$$
$$lk = b_{43} j + c_{43} k + d_{43} l$$
$$l^2 = b_4 j + c_4 k + d_4 l$$
so that there can be no pure algebra in this case if b_{41} vanishes,* and it may be assumed without loss of generality that
$$li = j.$$
There are two cases:
        [$232^2 1$], when d_3 does not vanish;
        [232^3], when d_3 vanishes.

[$232^2 1$]. The defining equation of this case can be reduced to
$$k^2 = l,$$
which gives
$$0 = k^3 = kl = lk = k^2 l = l^2,$$
and there is no pure algebra in this case.†

[232^3]. The defining equation of this case is
$$d_3 = 0,$$
which gives
$$0 = k^3 = c_3 k^2 = c_3, \quad k^2 = b_3 j,$$
$$0 = k^2 l = c_{34} k^2 + d_{34} kl = d_{34} = b_3 c_{34},$$
$$0 = lk^2 = c_{43} k^2 + d_{43} kl = d_{43} = b_3 c_{43}.$$
There are two cases:
        [$232^3 1$], when b_3 does not vanish;
        [232^4], when b_3 vanishes.

[$232^3 1$]. The defining equation of this case can be reduced to
$$k^2 = j,$$
which gives
$$0 = c_{34} = c_{43}, \quad kl = b_{34} j, \quad lk = b_{43} j,$$
$$k(l - b_{34} j) = 0,$$

---
* In this case, j, k and l, might form any one of the algebras (b_3), (c_3), (d_3) or (e_3). [C. S. P.]
† The case is impossible because $ki = 0$ and $k^2 i = j$. [C. S. P.]

so that $l - b_{34}j$ can be substituted for l without loss of generality, which is the same as to assume
$$kl = 0;$$
and this gives
$$0 = l^3 = d_4 l^2 = d_4 = c_4 lk = c_4 b_{43} = l^2 k = c_4,$$
so that there is no pure algebra in this case.*

[232^4]. The defining equation of this case is
$$k^2 = 0,$$
which gives
$$0 = lj = l^2 i = d_4 j = d_4,$$
$$0 = kl^2 = c_{34} kl = c_{34}, \quad kl = b_{34} j,$$
$$0 = l^2 k = c_{43} lk = c_{43}, \quad lk = b_{43} j,$$
and there can be no pure algebra if c_4 vanishes, so that it may be assumed, without loss of generality, that
$$l^2 = k,$$
which gives
$$0 = l^3 = lk = kl.$$

This gives a quadruple algebra which may be called (r_4), its multiplication table being †

(r_4)	i	j	k	l
i	j	0	0	0
j	0	0	0	0
k	0	0	0	0
l	j	0	0	k

[24]. The defining equations of this case are
$$i^2 = j^2 = k^2 = l^2 = 0,$$

---

* Substituting $i - l$ for i, this case is, $i = B:D, j = A:D, k = A:C + C:D, l = A:B$. [C. S. P.]

† $i = A:B + B:D + C:D, j = A:D, k = A:E, l = A:C + C:E$. [C. S. P.]

and it may be assumed, from §§ 63 and 65, that
$$ij = k = -ji, \quad il = li = 0,$$
which give
$$0 = ik = ki = jk = kj = kl = lk,$$
$$0 = ijl = b_{21}k = b_{24} = j^2l = -a_{21}k + d_{21}jl = d_{24} = a_{21},$$
$$jl = -lj = c_{21}k,$$
so that there is no pure algebra in this case.*

QUINTUPLE ALGEBRA.

There are two cases:

[1], when there is an idempotent basis;
[2], when the algebra is nilpotent.

[1]. The defining equation of this case is
$$i^2 = i.$$
There are eleven cases:

[1²], when j, k, l and m are all in the first group;
[12], when j, k and l are in the first, and m in the second group;
[13], when j and k are in the first, and l and m in the second group;
[14], when j and k are in the first, l in the second, and m in the third group;
[15], when j is in the first, and k, l and m in the second group;
[16], when j is in the first, k and l in the second, and m in the third group;
[17], when j is in the first, k in the second, l in the third, and m in the fourth group;
[18], when j, k and l are in the second, and m in the third group;
[19], when j and k are in the second, and l and m in the third group;
[10¹], when j and k are in the second, l in the third, and m in the fourth group;
[11¹], when j is in the second, k in the third, and l and m in the fourth group.

[1²]. The defining equations of this case are
$$ij = ji = j, \quad ik = ki = k, \quad il = li = l, \quad im = mi = m.$$
The algebras deduced by §69 from algebras (i_4) to (r_4) may be named (a_5) to (j_5), and their multiplication tables are respectively

---
*$i = -A:C + B:E$, $j = A:B + C:E + cD:E$, $k = A:E$, $l = -A:D + cB:E$. [C. S. P.]

(a_5)	i	j	k	l	m
i	i	j	k	l	m
j	j	k	l	m	0
k	k	l	m	0	0
l	l	m	0	0	0
m	m	0	0	0	0

(b_5)	i	j	k	l	m
i	i	j	k	l	m
j	j	k	l	0	0
k	k	l	0	0	0
l	l	0	0	0	0
m	m	l	0	0	l

(c_5)	i	j	k	l	m
i	i	j	k	l	m
j	j	k	l	0	0
k	k	l	0	0	0
l	l	0	0	0	0
m	m	l	0	0	0

(d_5)	i	j	k	l	m
i	i	j	k	l	m
j	j	k	0	0	l
k	k	0	0	0	0
l	l	0	0	0	0
m	m	$ak+bl$	0	0	l

or

(e_5)	i	j	k	l	m
i	i	j	k	l	m
j	j	k	0	0	l
k	k	0	0	0	0
l	l	0	0	0	0
m	m	$k-l$	0	0	0

(e_5)	i	j	k	l	m
i	i	j	k	l	m
j	j	0	0	0	l
k	k	0	0	0	0
l	l	0	0	0	0
m	m	k	0	0	0

(f_5)	i	j	k	l	m
i	i	j	k	l	m
j	j	k	0	0	l
k	k	0	0	0	0
l	l	0	0	0	0
m	m	al	0	0	0

(g_5)	i	j	k	l	m
i	i	j	k	l	m
j	j	k	0	0	0
k	k	0	0	0	0
l	l	0	0	0	0
m	$n.$	l	0	0	$l+ak$

(h_5)	i	j	k	l	m
i	i	j	k	l	m
j	j	k	0	0	0
k	k	0	0	0	0
l	l	0	0	0	0
m	m	l	0	0	k

(i_5)	i	j	k	l	m
i	i	j	k	l	m
j	j	k	0	0	0
k	k	0	0	0	0
l	l	0	0	0	0
m	m	l	0	0	0

(j_5)	i	j	k	l	m
i	i	j	k	l	m
j	j	k	0	0	0
k	k	0	0	0	0
l	l	0	0	0	0
m	m	k	0	0	l

[12]. The defining equations of this case are

$$ij = ji = j, \quad ik = ki = k, \quad il = li = l, \quad im = m, \quad mi = 0,$$

which give, by § 46,

$$0 = mj = mk = ml = m^2,$$

and if A is any expression belonging to the first group, but not involving i, we have the form

$$Am = am,$$

and by § 67, A is nilpotent, so that there is some power n which gives

$$0 = A^n = A^n m = aA^{n-1}m = a^n m = a = Am,$$
$$0 = jm = km = lm;$$

and there is no pure algebra in this case.*

[13]. The defining equations of this case are

$$ij = ji = j, \quad ik = ki = k, \quad il = l, \quad im = m, \quad li = mi = 0,$$

which give, by § 46,

$$0 = lj = lk = l^2 = lm = mj = mk = ml = m^2;$$

and it may be assumed from (a_3), by § 69, that

$$j^2 = k, \quad j^3 = 0.$$

It may also be assumed that

$$jl = m, \quad \text{whence} \dagger \quad kl = jm = 0.$$

We thus obtain a quintuple algebra which may be called (k_5), its multiplication table being this: ‡

---

* In fact i and m, by themselves, form the algebra (b_2), while i, j, k, l, by themselves form one of the algebras (a_4), (b_4), (c_4), (d_4), the products of m with j, k and l vanishing. [C. S. P.]

† This is proved as follows: $0 = j^2 l = j^2 m = d_{25} jl + e_{25} jm = d_{25} e_{25} l + (d_{25} + e_{25}^2)m$. Thus $d_{25} e_{25} = 0$ and $d_{25} + e_{25}^2 = 0$; or $d_{25} = 0$, $e_{25} = 0$, $jm = kl = 0$. [C. S. P.]

‡ $i = A:A + B:B + C:C$, $j = A:B + B:C$, $k = A:C$, $l = B:D$, $m = A:D$. [C. S. P.]

(k_5)	i	j	k	l	m
i	i	j	k	l	m
j	j	k	0	m	0
k	k	0	0	0	0
l	0	0	0	0	0
m	0	0	0	0	0

[14]. The defining equations of this case are

$$ij = ji = j, \quad ik = ki = k, \quad il = l, \quad mi = m, \quad li = im = 0,$$

which give, by § 46,

$$0 = jm = km = lj = lk = l^2 = ml = m^2.$$

It may be assumed from § 69 and (a_3) that

$$j^2 = k, \quad j^3 = 0,$$

whence

$$0 = jl = kl = mj = mk = jlm = a_{45}j + b_{45}k = a_{45} = b_{45}, \quad lm = c_{45}k,$$

and there is no pure algebra in this case.*

[15]. The defining equations of this case are

$$ij = ji = j, \quad ik = k, \quad il = l, \quad im = m, \quad ki = li = mi = 0,$$

which give, by §§ 46 and 67,

$$0 = j^2 = kj = k^2 = kl = km = lj = lk = l^2 = lm = mj = mk = ml = m^2.$$

It may be assumed that $\quad jk = l, \quad jm = 0,$†

whence, $\qquad\qquad jl = 0,$

and there is no pure algebra in this case.‡

---

*$i = A:A+B:B+C:C, j = A:B+B:C, k = A:C, l = A:D, m = cD:C.$ [C. S. P.]

† We cannot suppose $jk = k$, because $j^2k = 0$. We may, therefore, put l for jk. Then $jl = 0$. Then, $0 = j^2m = c_{25}e_{25}k + (d_{25}c_{25} + c_{25})l + e_{25}^2m$. It follows that $jm = d_{25}l$, and substituting $m - d_{25}k$ for m, we have $jm = 0$. The algebra thus separates into (b_2) and (e_4). [C. S. P.]

‡ $i = A:A+B:B, j = A:B, k = B:C, l = A:C, m = A:D.$ [C. S. P.]

[16]. The defining equations of this case are
$$ij = ji = j, \quad ik = k, \quad il = l, \quad mi = m, \quad ki = li = im = 0,$$
which give, by §§ 46 and 67,
$$0 = j^2 = jm = kj = k^2 = kl = lj = lk = l^2 = mj = mk = ml = m^2,$$
$$km = a_{35}i + b_{35}j, \quad lm = a_{45}i + b_{45}j,$$
and it may be assumed that
$$jk = d_{23}l, \quad jl = 0,$$
and d_{23} cannot vanish in the case of a pure algebra,* so that it is no loss of generality to assume
$$jk = l,$$
which gives
$$jkm = lm = a_{35}j.$$
There are two cases:
[161], when a_{35} does not vanish;
[162], when a_{35} vanishes.

[161]. The defining equation of this case can be reduced to
$$a_{35} = 1,$$
which gives
$$lm = j, \quad km = i + b_{35}j,$$
and $i + b_{35}j$ can be substituted for i, and this gives a quintuple algebra which may be called (l_5), of which the multiplication table is

(l_5)	i	j	k	l	m
i	i	j	k	l	0
j	i	0	l	0	0
k	0	0	0	0	i
l	0	0	0	0	j
m	m	0	0	0	0

* But $0 = mk = kmk = (a_{35}i + b_{35}j)k = a_{35}k + d_{23}b_{35}l$. Hence $a_{35} = 0$ and either d_{23} or $b_{35} = 0$, and in either case there is no pure algebra. The two algebras (l_5) and (m_5) are incorrect, as may be seen by comparing $k \cdot mk$ with $km \cdot k$. [C. S. P.]

[162]. The defining equation of this case is

$$a_{35} = 0,$$

which gives

$$km = b_{35}j, \quad lm = 0;$$

and b_{35} cannot vanish in the case of a pure algebra, so that it is no loss of generality to assume

$$km = j.$$

This gives a quintuple algebra which may be called (m_5), of which the multiplication table is

(m_5)	i	j	k	l	m
i	i	j	k	l	0
j	j	0	l	0	0
k	0	0	0	0	j
l	0	0	0	0	0
m	m	0	0	0	0

[17]. The defining equations of this case are

$$ij = ji = j, \quad ik = k, \quad li = l, \quad ki = il = im = mi = 0,$$

which give, by §§ 46 and 67,

$$0 = j^2 = jk = jl = jm = kj = k^2 = lj = l^2 = lm = mj = mk,$$
$$kl = a_{34}i + b_{34}j, \quad km = c_{35}k, \quad lk = c_{43}m, \quad ml = d_{54}l, \quad m^2 = e_5 m,$$
$$0 = jkl = a_{34}j = a_{34},$$
$$lkl = b_{34}lj = 0 = c_{43}ml = c_{43}d_{54}, \quad klk = b_{34}jk = 0 = c_{43}km = c_{43}c_{35},$$
$$lkm = c_{35}lk = c_{43}m^2 = c_{35}c_{43}m = 0 = c_{43}e_5,$$
$$kml = d_{54}kl = c_{35}kl, \quad (d_{54} - c_{35})b_{34} = 0, \quad km^2 = e_5 km = c_{35}km, \quad (e_5 - c_{35})c_{35} = 0,$$
$$m^2 l = e_5 ml = d_{54}ml, \quad (e_5 - d_{54})d_{54} = 0.$$

There are two cases:

[171], when $e_5 = 1$;*
[172], when $e_6 = 0$.

[171]. The defining equation of this case is
$$m^2 = m,$$
which gives
$$0 = c_{43} = lk.$$

There can be no pure algebra if either of the quantities b_{34}, c_{33} or d_{54} vanish, and there is no loss of generality in assuming
$$kl = j, \quad km = k, \quad ml = l.$$

This gives a quintuple algebra which may be called (n_5), its multiplication table being

(n_5)	i	j	k	l	m
i	i	j	k	0	0
j	j	0	0	0	0
k	0	0	0	j	k
l	l	0	0	0	0
m	0	0	0	l	m

[172]. The defining equation of this case is
$$m^2 = 0,$$
which gives
$$0 = c_{35} = d_{34} = km = ml;$$

---

* But on examination of the assumptions already made, it will be seen that if e_5 is not zero, and consequently $c_{43} = 0$, the algebra breaks up into two. Accordingly, the algebra (n_5) is impure, for i, j, k and l, alone, form the algebra (f_4), while m, l, k, j, alone, form the algebra (h_4), and $im = mi = 0$. [C. S. P.]

and there can be no pure algebra if either b_{34} or c_{43} vanishes, and it may be assumed that
$$kl = j, \quad lk = m.$$
This gives a quintuple algebra which may be called (o_5), its multiplication table being as follows:*

(o_5)	i	j	k	l	m
i	i	j	k	0	0
j	j	0	0	0	0
k	0	0	0	j	0
l	l	0	m	0	0
m	0	0	0	0	0

[18]. The defining equations of this case are
$$ij = j, \quad ik = k, \quad il = l, \quad mi = m, \quad ji = ki = li = im = 0,$$
which give, by § 46,
$$0 = j^2 = jk = jl = kj = k^2 = kl = lj = lk = l^2 = mj = mk = ml = m^2.$$
But if A is any expression of the second group,
$$Am = ai;$$
which gives
$$0 = Amj = aj = a = Am = jm = km = lm,$$
and there is no pure algebra in this case.

[19]. The defining equations of this case are
$$ij = j, \quad ik = k, \quad li = l, \quad mi = m, \quad il = im = ji = ki = 0,$$
which give, by § 46,
$$0 = j^2 = jk = kj = k^2 = lj = lk = l^2 = lm = mj = mk = ml = m^2.$$

---

*$i = B:B + D:D + F:F$, $j = D:F$, $k = B:C + D:E$, $l = A:B + E:F$, $m = A:C$. [C. S. P.]

But if A is an expression of the second group and B one of the third,
$$AB = ai,$$
which gives
$$0 = ABj = aj = a = AB = jl = jm = kl = lm,$$
and there is no pure algebra in this case.

[10′]. The defining equations of this case are
$$ij = j, \quad ik = k, \quad li = l, \quad ji = ki = il = im = mi = 0,$$
which give, by § 46,
$$0 = j^2 = jk = kj = k^2 = l^2 = lm = mj = mk;$$
and it is obvious that we may assume
$$jl = 0.$$
We have, then,
$$jm = b_{25}j + c_{25}k, \quad kl = a_{34}i, \quad km = b_{35}j + c_{35}k,$$
$$lj = e_{42}m, \quad lk = e_{43}m, \quad ml = d_{54}l, \quad m^2 = e_5 m,$$
$$0 = d_{54}jl = jml = c_{25}kl = a_{34}c_{25}i = a_{34}c_{25}.$$

There are two cases :
              [10′1], when a_{34} does not vanish ;
              [10′2], when a_{34} vanishes.

[10′1]. The defining equation of this case can be reduced to
$$kl = i,$$
which gives
$$c_{25} = 0, \quad jm = b_{25}j.$$
There are two cases:
              [10′1²], when $e_5 = 1$;
              [10′12], when e_5 vanishes.

[10′1²]. The defining equation of this case is
$$m^2 = m;$$
and we assume
$$jm = j, \quad ml = l, \quad km = k,$$
because otherwise this case would coincide with a subsequent one. We get, then,
$$0 = jlj = e_{42}jm = e_{42} = lj, \quad 0 = jlk = e_{43}jm = e_{43} = lk,$$
which virtually brings this case under [10′2].*

---

    * This does not seem clear. But $i = i^2 = klkl = 0$, which is absurd. [C. S. P.]

[10'12]. The defining equation of this case is
$$m^2 = 0,$$
which gives
$$0 = jm^2 = b_{25}jm = b_{25} = jm, \quad 0 = m^2l = d_{52}ml = a_{54} = ml,$$
$$0 = km^2 = c_{35}km = c_{35}, \quad km = b_{35}j, \quad lkl = li = l = c_{43}ml = 0,$$
which is impossible, and this case disappears.

[10'2]. The defining equation of this case is
$$kl = 0.*$$
There are two cases:
[10'21], when $e_5 = 1$;
[10'2²], when e_5 vanishes.

[10'21]. The defining equation of this case is
$$m^2 = m,$$
and if we would not virtually proceed to a subsequent case, we must assume
$$jm = j, \quad km = k, \quad ml = l,$$
and there is no loss of generality in assuming
$$lj = 0,$$
so that there is no pure algebra in this case.†

[10'2²]. The defining equation of this case is
$$m^2 = 0,$$
which gives
$$0 = m^2l = d_{54}ml = d_{54} = ml;$$
and we may assume
$$c_{25} = 0,$$
which gives
$$0 = jm^2 = b_{25}jm = b_{25} = jm, \quad 0 = km^2 = c_{35}km = c_{35}, \quad km = b_{35}j,$$
$$0 = c_{43}m^2 = lkm = b_{35}c_{42}m = b_{35}c_{42};‡$$

---

* In this case, the algebra at once separates into an algebra between j, k, l and m, and three double algebras between i and j, i and k, and i and l, respectively. [C. S. P.]

† In fact, $0 = lklk = e_4^2 m = e_{41} = lk$. So that the algebra falls into six parts of the form (b_2). [C. S. P.]

‡ The author omits to notice that $0 = klk = e_{41}km = e_{41}b_{35}$. Thus, either $km = 0$ or $lj = lk = 0$. The algebra (p_5) involves an inconsistency in regard to klk. [C. S. P.]

and we have without loss of generality
$$lj = 0, \quad km = j, \quad lk = m.$$
This gives a quintuple algebra which may be called (p_5), of which the multiplication table is

(p_5)	i	j	k	l	m
i	i	j	k	0	0
j	0	0	0	0	0
k	0	0	0	0	j
l	l	0	m	0	0
m	0	0	0	0	0

[11′]. The defining equations of this case are
$$ij = j, \quad ki = k, \quad ji = ik = il = im = li = mi = 0;$$
which give, by § 46,
$$0 = j^2 = k^2 = kl = km = lj = mj,$$
$$jk = a_{23}i, \quad jl = b_{24}j, \quad jm = b_{25}j, \quad kj = d_{32}l + e_{32}m, \quad lk = c_{43}k, \quad mk = c_{53}k.$$
There are two cases:

[11′1], when l is the idempotent base of the fourth group;
[11′2], when the fourth group is nilpotent.

[11′1]. The defining equation of this case is
$$l^2 = l.$$
There are two cases:

[11′1²], when m is in the second subsidiary group of the fourth group;
[11′12], when m is in the fourth subsidiary group of the fourth group.

[11′1²]. The defining equations of this case are
$$lm = m, \quad ml = 0;$$

PEIRCE: *Linear Associative Algebra.* 59

which give
$$0 = m^2 = jm^2 = b_{25}jm = b_{25} = jm,$$
$$0 = m^2k = c_{53}mk = c_{53} = mk;$$

and a_{23} cannot vanish in a pure algebra, so that we may assume
$$jk = i,$$
which gives
$$kjk = k = d_{32}c_{43}k, \quad jkj = j = d_{32}b_{24}j, \quad 1 = d_{32}c_{43} = d_{32}b_{24},$$
$$jl = jl^2 = b_{24}jl, \quad b_{24}^2 = b_{24} = 1, \quad lk = l^2k = c_{43}lk, \quad c_{43}^2 = c_{43} = 1 = d_{32},$$
$$jl = j, \quad lk = k, \quad kjl = l = kj,$$

and there is no pure algebra in this case.*

[11'12]. The defining equations of this case are
$$lm = ml = 0,$$
which give
$$0 = jlm = b_{24}jm = b_{24}b_{25}j = b_{24}b_{25}, \quad 0 = lmk = c_{53}lk = c_{43}c_{53}k = c_{43}c_{53},$$
$$kjl = d_{32}l = b_{24}kj = b_{24}d_{32}l + b_{24}c_{32}m, \quad lkj = d_{32}l = c_{43}kj = c_{43}d_{32}l + c_{43}c_{32}m,$$
$$kjm = c_{32}m^2 = b_{25}kj = b_{25}d_{32}l + b_{25}c_{32}m, \quad mkj = c_{32}m^2 = c_{53}kj = c_{53}d_{32}l + c_{53}c_{32}m,$$
$$d_{32} = b_{24}d_{32} = c_{43}d_{32}, \quad 0 = b_{25}d_{32} = c_{53}d_{32} = b_{24}c_{32} = c_{43}c_{32}.\dagger$$

There are two cases:

[11'121], when m is idempotent;
[11'12²], when m is nilpotent.

[11'121]. The defining equation of this case is
$$m^2 = m,$$
which gives
$$c_{32} = c_{53}e_{32} = b_{25}e_{32};$$
and it may be assumed that
$$b_{24} = 0.$$

But if the algebra is then regarded as having l for its idempotent basis, it is evident from § 50 that the bonds required for a pure algebra are wanting, so that there is no pure algebra in this case.‡

---

*In fact, i, j, k, l form the algebra (g_4), and l, m, the algebra (b_2). [C. S. P.]
†The last equation holds by § 68. [C. S. P.]
‡Namely, $d_{32} = 0$, and either $e_{32} = 1$, when l forms the algebra (a_1), and i, j, k, m the algebra (g_4), or else $e_{32} = 0$, when by [18] of triple algebra $a_{23} = 0$, and j and k each forms the algebra (b_2) with each of the letters i, l, m. [C. S. P.]

[11'12²]. The defining equation of this case is
$$m^2 = 0,$$
which gives
$$0 = jm^2 = b_{25}jm = b_{25}^2 j = b_{25} = jm, \quad 0 = m^2 k = c_{53}mk = c_{53}^2 k = c_{53} = mk,$$
$$1 = b_{24} = c_{43}, \quad jl = j, \quad lk = k, \quad 0 = e_{32},$$
and there is no pure algebra in this case.*

[11'2]. The defining equation of this case is
$$l^n = 0,$$
in which n is 2 or 3. We must then have
$$0 = lm = ml = m^2,$$
which give
$$0 = jl^3 = b_{24}jl^2 = b_{24}^2 jl = b_{24} = jl = jm = lk = mk, \quad 0 = kjk = a_{23}k = a_{23} = jk,$$
and there is no pure algebra in this case. †

[2]. The defining equation of this case is
$$i^n = 0.$$
There are five cases:

[21], when $n = 6$;
[2²], when $n = 5$;
[23], when $n = 4$;
[24], when $n = 3$;
[25], when $n = 2$.

[21]. The defining equation of this case is
$$i^6 = 0,$$
and by § 60,
$$i^2 = j, \quad i^3 = k, \quad i^4 = l, \quad i^5 = m.$$

This gives a quintuple algebra which may be called (q_5), its multiplication table being

---

* Here, m forms the algebra (b_1), and the other letters form (g_4). [C. S. P.]

† Namely, if $n = 2$, j, l, k, form the algebra (d_3) (second form). i, j, and i, k. the algebra (b_2), and m the algebra (c_1). But if $n = 3$. j. k. l and m form an algebra transformable into (j_4) or (k_4), while i, j. and i. k form. each pair. the algebra (b_2). [C. S. P.]

(q_5)	i	j	k	l	m
i	j	k	l	m	0
j	k	l	m	0	0
k	l	m	0	0	0
l	m	0	0	0	0
m	0	0	0	0	0

[2⁹]. The defining equation of this case is

$$i^5 = 0,$$

and by § 59,

$$i^2 = j, \quad i^3 = k, \quad i^4 = l.$$

There are then by § 64 two quintuple algebras which may be called (r_5) and (s_5), their multiplication tables being

(r_5)	i	j	k	l	m
i	j	k	l	0	0
j	k	l	0	0	0
k	l	0	0	0	0
l	0	0	0	0	0
m	l	0	0	0	l

(s_5)	i	j	k	l	m
i	j	k	l	0	0
j	k	l	0	0	0
k	l	0	0	0	0
l	0	0	0	0	0
m	l	0	0	0	0

[23]. The defining equation of this case is

$$i^4 = 0;$$

and by § 59,

$$i^2 = j, \quad i^3 = k;$$

and it may be assumed, from the principle of § 63, that
$$il = 0,$$
which gives
$$0 = jl = kl = ili = il^2 = ilm$$
$$li = c_{41}k + d_{41}l + e_{41}m, \quad l^2 = c_4k + d_4l + e_4m, \quad lm = c_{45}k + d_{45}l + e_{45}m.$$

There are two cases:
[231], when $im = l$;
[232], when $im = 0$.

[231]. The defining equation of this case is
$$im = l,$$
whence
$$0 = jm = km = jmi = jml = jm^2 = e_{41} = e_4 = e_{45},$$
$$li^2 = d_4li, \quad 0 = li^4 = d_{41}li^3 = d_{41}^2li^2 = d_{41}^3li = d_{41} = lj = lk,$$
$$l^3 = d_4l^2, \quad 0 = l^4 = d_4l^3 = d_4, \quad li = c_{41}k, \quad l^2 = c_4k, \quad lm = c_{45}k + d_{45}l,$$
$$imi = li = c_{41}k, \quad mi = c_{41}j + c_{51}k + d_{51}l, \quad mj = c_{41}(1 + d_{51})k, \quad mk = 0,$$
$$iml = l^2 = c_4k, \quad ml = c_4j + c_{54}k + d_{54}l,$$
$$im^2 = lm = c_{45}k + d_{45}l, \quad m^2 = c_{45}j + c_5k + d_5l + d_{45}m,$$
$$0 = m^4 = d_{45}, \quad lim = l^2 = c_{41}km = 0 = mli = d_{54}c_{41}k = d_{54}c_{41},$$
$$0 = mlm = d_{54}lm = d_{54}c_{45}, \quad 0 = m^2l = d_{54}ml = d_{54}.^*$$

There are two cases:
[231²], when c_{41} does not vanish;
[2312], when c_{41} vanishes.

[231²]. The defining equation of this case is reducible to
$$li = k.$$
There are two cases:
[231³], when c_{45} does not vanish;
[231²2], when c_{45} vanishes.

[231³]. The defining equation of this case can be reduced to
$$lm = k,$$
which gives
$$m^2i = k + d_{51}k + d_{51}^2k = k + d_5k, \quad d_5 = d_{51} + d_{51}^2,$$
$$m^3 = k + d_{51}k + d_5d_{51}k = d_5k, \quad d_{51}^3 = -1;$$

---

* To these equations are to be added the following, which is taken for granted below: $ml = mim = c_{44}d_{51}k$. [C. S. P.]

and if $\mathfrak{r}$ is one of the imaginary cube roots of -1, there are two cases:

$$[231^4], \text{ when } d_{51} = \mathfrak{r};$$
$$[231^32], \text{ when } d_{51} = -1.$$

[231⁴]. The defining equation of this case is

$$d_{51} = \mathfrak{r},$$

which gives

$$i(m - c_{51}l) = l, \quad l(m - c_{51}l) = k,$$
$$(m - c_{51}l)i = j + \mathfrak{r}l, \quad (m - c_{51}l)j = (1 + \mathfrak{r})k,$$
$$(m - c_{51}l)k = 0, \quad (m - c_{51}l)l = \mathfrak{r}k,$$
$$(m - c_{51}l)^2 = j + [c_5 - c_{51}(1 + \mathfrak{r})]k + (2\mathfrak{r} - 1)l;$$

so that the substitution of $m - c_{51}l$ for m is the same as to make

$$c_{51} = 0.$$

There are two cases:

$$[231^5], \text{ when } c_5 \text{ does not vanish};$$
$$[231^42], \text{ when } c_5 \text{ vanishes}.$$

[231⁵]. The defining equation of this case can be reduced to

$$c_5 = 1.$$

There is then a quintuple algebra which may be called (t_5), its multiplication table being*

---

* The author has overlooked the circumstance that (t_5) and (u_5) are forms of the same algebra. If in (t_5) we put $i_1 = i - \mathfrak{r}^2 j$, $j_1 = j - 2\mathfrak{r}^2 k$, $k_1 = k$, $l_1 = -\mathfrak{r}^2 k + l$, $m_1 = -\mathfrak{r}^2 j + m$, we get (u_5). The structure of this algebra may be shown by putting $i_1 = \mathfrak{r}i$, $j_1 = \mathfrak{r}^2 j$, $k_1 = -k$, $l_1 = \mathfrak{r}^2 j - \mathfrak{r}l$, $m_1 = \mathfrak{r}i - m$, when we have this multiplication table (where the subscripts are dropped):

(u_5)	i	j	k	l	m
i	j	k	0	k	l
j	k	0	0	0	k
k	0	0	0	0	0
l	$\mathfrak{r}k$	0	0	0	0
m	$\mathfrak{r}l$	$\mathfrak{r}^2 k$	0	0	0

In relative form, $i = A:B + A:C + B:E + C:D + E:G$, $j = A:D + A:E + B:G$, $k = A:G$, $l = \mathfrak{r}A:E + C:G$, $m = \mathfrak{r}^2 A:B + A:F + \mathfrak{r}C:E + D:G - F:G$. [C. S. P.]

(t_5)	i	j	k	l	m
i	j	k	0	0	l
j	k	0	0	0	0
k	0	0	0	0	0
l	k	0	0	0	k
m	$j+\mathfrak{r}l$	$(1+\mathfrak{r})k$	0	$\mathfrak{r}k$	$j+k+(2\mathfrak{r}-1)l$

[231^{4}2]. The defining equation of this case is
$$c_5 = 0.$$
There is then a quintuple algebra which may be called (u_5), its multiplication table being

(u_5)	i	j	k	l	m
i	j	k	0	0	l
j	k	0	0	0	0
k	0	0	0	0	0
l	k	0	0	0	k
m	$j+\mathfrak{r}l$	$(1+\mathfrak{r})k$	0	$\mathfrak{r}k$	$j+(2\mathfrak{r}-1)l$

[231^{3}2]. The defining equation of this case is
$$d_{51} = -1,$$
which gives
$$d_5 = 0, \quad i(m - c_{51}l) = l, \quad l(m - c_{51}l) = k,$$
$$(m - c_{51}l)i = j - l, \quad (m - c_{51}l)l = -k, \quad (m - c_{51}l)_2 = j + c_5k;$$
so that the substitution of $m - c_{51}l$* for m is the same as to make
$$c_{51} = 0.$$

---

* The original text has $m - c_{51}k$ throughout these equations, but it is plain that $m - c_{51}l$ is meant. [C. S. P.]

There are two cases:
[231²21], when c_5 does not vanish;
[231²2²], when c_5 vanishes.

[231²21]. The defining equation of this case can be reduced to

$$c_5 = 1.$$

There is a quintuple algebra which may be called (v_5), its multiplication table being*

(v_5)	i	j	k	l	m
i	j	k	0	0	l
j	k	0	0	0	0
k	0	0	0	0	0
l	k	0	0	0	k
m	$j-l$	0	0	$-k$	$j+k$

[231²2²]. The defining equation of this case is

$$c_5 = 0.$$

This gives a quintuple algebra which may be called (w_5), its multiplication table being*

---

* The algebra (v_5) reduces to (w_5) on substituting $i_1 = i + \frac{1}{3}j + \frac{1}{3}l$, $j_1 = j+k$, $k_1 = k$, $l_1 = \frac{2}{3}k + l$, $m_1 = \frac{1}{3}j + \frac{1}{3}l + m$. To exhibit the structure of this algebra, we may put ρ and ρ' for imaginary cube roots of 1, and substitute in (w_5) $i_1 = i + \rho'm$, $j_1 = (1-\rho)j + k + \sqrt{-3}l$, $k_1 = 3k$, $l_1 = (1-\rho')j + k - \sqrt{-3}l$, $m_1 = i + \rho m$. Then, dropping the subscripts, we have this multiplication table.

	i	j	k	l	m
i	0	0	0	k	j
j	k	0	0	0	0
k	0	0	0	0	0
l	0	0	0	0	k
m	l	k	0	0	0

In relative form. $i = \rho'A:B + \rho'C:F + 3\rho D:E$, $j = 3\rho A:C + 3\rho'D:F$, $k = 3A:D$, $l = 3\rho'A:E + 3\rho B:F$. $m = \rho A:D + 3\rho'B:C + \rho E:F$. [C. S. P.]

(w)	i	j	k	l	m
i	j	k	0	0	l
j	k	0	0	0	0
k	0	0	0	0	0
l	k	0	0	0	k
m	$j-l$	0	0	$-k$	$j+k$

[$231^2 2$]. The defining equation of this case is
$$lm = 0,$$
which gives
$$ml = 0, \quad m^2 = c_5 k + d_5 l, \quad m^2 i = d_5 k = [1 + d_{51}] k, \quad d_5 = 1 + d_{51},$$
and c_{51} may be made to vanish without loss of generality.

There are three cases:

[$231^2 21$], when neither d_{51} nor $d_{51} + 1$ vanishes;
[$231^2 2^2$], when $d_{51} + 1$ vanishes;
[$231^2 23$], when d_{51} vanishes.

[$231^2 21$]. The defining formulae of this case are
$$d_{51} \neq 0, \quad d_{51} \neq -1.$$
There are two cases:

[$231^2 21^2$], when c_5 does not vanish;
[$231^2 212$], when c_5 vanishes.

[$231^2 21^2$]. The defining equation of this case can always be reduced to
$$c_5 = 1.$$
This gives a quintuple algebra which may be called (x_5), its multiplication table being*

---

* In relative form, $i = A:B + A:E + B:D + D:F$, $j = A:D + B:F$, $k = A:F$, $l = A:D$, $m = (1+a) A:B + A:C + A:E + B:D + C:D + D:F + E:F$. [C. S. P.]

(x_5)	i	j	k	l	m
i	j	k	0	0	l
j	k	0	0	0	0
k	0	0	0	0	0
l	k	0	0	0	0
m	$j+al$	$(1+a)k$	0	0	$k+(1+a)l$

[231^2212]. The defining equation of this case is

$$c_5 = 0.$$

This gives a quintuple algebra which may be called (y_5), its multiplication table being*

(y_5)	i	j	k	l	m
$i.$	j	k	0	0	l
j	k	0	0	0	0
k	0	0	0	0	0
l	k	0	0	0	0
m	$j+ml$	$(1+a)k$	0	0	$(1+a)l$

[231^22^2]. The defining equation of this case is

$$d_{51} = -1,$$

which gives

$$mi = j - l, \quad mj = 0, \quad m^2 = c_5k.$$

There are two cases:

[231^22^21], when c_5 does not vanish;
[231^22^3], when c_5 vanishes.

---

* The relative form is the same as that of (x_5): omitting from m the terms $A:E$ and $E:F$. [C. S. P.]

[$231^2 2^2 1$]. The defining equation of this case can be reduced to
$$m^2 = k.$$
This gives a quintuple algebra which may be called (z_5), its multiplication table being *

(z_5)	i	j	k	l	m
i	j	k	0	0	l
j	k	0	0	0	0
k	0	0	0	0	0
l	k	0	0	0	0
m	$j-l$	0	0	0	k

[$231^2 2^3$]. The defining equation of this case is
$$m^2 = 0.$$
This gives a quintuple algebra which may be called (aa_5), its multiplication table being †

(aa_5)	i	j	k	l	m
i	j	k	0	0	l
j	k	0	0	0	0
k	0	0	0	0	0
l	k	0	0	0	0
m	$j-l$	0	0	0	0

---

*In relative form, $i = A:B + B:C + C:D$, $j = A:C + B:D$, $k = A:D$. $l = A:C$, $m = B:C + A:E + E:D$. [C. S. P.]

† In relative form, the same as (z_5), except that $m = B:C$. [C. S. P.]

[$23l^223$]. The defining equation of this case is
$$mi = j,$$
which gives
$$0 = (l-j)i = (m-i)i;$$
so that, by the substitution of $l-j$ for l and $m-i$ for m, this case would virtually be reduced to [232].

[2312]. The defining equation of this case is
$$li = 0,$$
which gives
$$mj = 0, \quad mim = ml = d_{51}lm, \quad d_{45} = 0, \quad c_{54} = d_{51}c_{45},$$
$$m^2i = d_{51}ml = c_{45}k, \quad c_{45} = d_{51}c_{54}, \quad m^3 = d_5lm = d_5ml, \quad d_5(c_{54} - c_{45}) = 0.$$

There are two cases:

[23121], when d_5 does not vanish;
[2312^2], when d_5 vanishes.

[23121]. The defining equation of this case can be reduced to
$$d_5 = 1,$$
which gives
$$c_{45} = c_{54};$$
and it may be assumed without loss of generality that
$$c_5 = 0.\text{*}$$

There are two cases:

[23121^2], when c_{45} does not vanish;
[231212], when c_{45} vanishes.

[23121^2]. The defining equation of this case can be reduced to
$$lm = ml = k,$$
which gives
$$d_{51} = 1.$$

There are two cases:

[23121^3], when c_{51} does not vanish;
[23121^22], when c_{51} vanishes.

[23121^3]. The defining equation of this case can be reduced to
$$c_{51} = 1.$$

---

* Namely, by putting $l_1 = c_5k + l$, $m_1 = m - c_5j$. [C. S. P.]

This gives a quintuple algebra which may be called (ab_5), its multiplication table being *

(ab_5)	i	j	k	l	m
i	j	k	0	0	l
j	k	0	0	0	0
k	0	0	0	0	0
l	0	0	0	0	k
m	$k+l$	0	0	k	$j+l$

[23121²2]. The defining equation of this case is

$$c_{51} = 0.$$

This gives a quintuple algebra which may be called (ac_5), its multiplication table being †

---

* The structure of this algebra is best seen on making the following substitutions: Let $\mathfrak{h}_1$ and $\mathfrak{h}_2$ represent the two roots of the equation $x^2 = x + 1$. That is, $\mathfrak{h}_1 = \frac{1}{2}(1 + \sqrt{5})$ and $\mathfrak{h}_2 = \frac{1}{2}(1 - \sqrt{5})$. Then substitute $i_1 = \mathfrak{h}_1^{-\frac{3}{2}}(i + \mathfrak{h}_1 m)$, $j_1 = \mathfrak{h}_1^{-\frac{3}{2}}\{(2+\mathfrak{h}_1)j + \mathfrak{h}_1 k + (1+3\mathfrak{h}_1)l\}$, $k_1 = \frac{1}{5}k$, $l_1 = \mathfrak{h}_2^{-\frac{3}{2}}\{(2+\mathfrak{h}_2)j + \mathfrak{h}_2 k + (1+3\mathfrak{h}_2)l\}$, $m_1 = \mathfrak{h}_2^{-\frac{3}{2}}(i + \mathfrak{h}_2 m)$. Then, we have the multiplication table:

	i	j	k	l	m
i	j	k	0	0	$\frac{1}{5}\mathfrak{h}_2 k$
j	k	0	0	0	0
k	0	0	0	0	0
l	0	0	0	0	k
m	$\frac{1}{5}\mathfrak{h}_1 k$	0	0	k	l

In relative form, $i = A:B + B:C + C:D + \frac{1}{5}\mathfrak{h}_1 A:G + H:D$, $j = A:C + B:D$, $k = A:D$, $l = A:F + E:D$, $m = A:E + E:F + F:D + \frac{1}{5}\mathfrak{h}_2 A:H + G:D$. [C. S. P.]

† On making the same substitutions for i and m as in the last note, this algebra falls apart into two algebras of the form (b_3). [C. S. P.]

(ac_5)	i	j	k	l	m
i	j	k	0	0	l
j	k	0	0	0	0
k	0	0	0	0	0
l	0	0	0	0	k
m	l	0	0	k	$j+l$

[231212]. The defining equation of this case is
$$ml = lm = 0.$$
There are two cases:

[2312121], when c_{51} does not vanish;
[231212²], when c_{51} vanishes.

[2312121]. The defining equation of this case can be reduced to
$$c_{51} = 1.$$
This gives a quintuple algebra which may be called (ad_5), its multiplication table being *

(ad_5)	i	j	k	l	m
i	j	k	0	0	l
j	k	0	0	0	0
k	0	0	0	0	0
l	0	0	0	0	0
m	$k+al$	0	0	0	l

---

* In relative form, $i = A:B + B:C + C:D + E:F + aF:G$, $j = A:C + B:D + aE:G$, $k = A:D$, $l = E:G$, $m = A:C + E:F + F:G$. [C. S. P.]

[231212^2]. The defining equation of this case is
$$c_{51} = 0.$$
This gives a quintuple algebra which may be called (ae_5), its multiplication table being *

(ae_5)	i	j	k	l	m
i	j	k	0	0	l
j	k	0	0	0	0
k	0	0	0	0	0
l	0	0	0	0	0
m	al	0	0	0	l

[2312^2]. The defining equation of this case is
$$d_5 = 0.$$
There are two cases:
  [2312^21], when c_{45} does not vanish;
  [2312^3], when c_{45} vanishes.

[2312^21]. The defining equation of this case can be reduced to
$$lm = k,$$
which gives
$$c_{45} = d_{51}^2 c_{45}, \quad d_{51}^2 = 1.$$
There are two cases:
  [2312^21^2], when $d_{51} = 1$;
  [2312^212], when $d_{51} = -1$.

[2312^21^2]. The defining equation of this case is
$$d_{51} = 1,$$
which gives
$$c_{51} = 1, \quad ml = k.$$

---
* In relative form, the same as (ad_5) except that $m = E:F + F:G$. [C. S. P.]

There are two cases:
$[2312^21^3]$, when c_{51} does not vanish;
$[2312^21^22]$, when c_{51} vanishes.

$[2312^21^3]$. The defining equation of this case can be reduced to

$$c_{51} = 1.$$

This gives a quintuple algebra which may be called (af_5), its multiplication table being*

(af_5)	i	j	k	l	m
i	j	k	0	0	l
j	k	0	0	0	0
k	0	0	0	0	0
l	0	0	0	0	k
m	$k+l$	0	0	k	$j+ck$

---

*To show the construction of this algebra, we may substitute $i_1 = i + m$, $j_1 = 2j + (a+1)k + 2l$, $k_1 = 4k$, $l_1 = 2j + (a-1)k - 2l$, $m_1 = i - m$. This gives the following multiplication table:

	i	j	k	l	m
i	j	k	0	0	$-\frac{a-1}{4}k$
j	k	0	0	0	0
k	0	0	0	0	0
l	0	0	0	0	k
m	$-\frac{a+1}{4}k$	0	0	k	l

This algebra thus strongly resembles (ab_5). In relative form, $i = A:B + B:C + C:D + A:G - \frac{a+1}{4} G:D$, $j = A:C + B:D - \frac{a+1}{4} A:D$, $k = A:D$, $l = A:F + E:D - \frac{a-1}{4} A:D$, $m = A:E + E:F + F:D + A:G - \frac{a-1}{4} G:D$. [C. S. P.]

[2312^21^22]. The defining equation of this case is
$$c_{51} = 0.$$
There are two cases:

[2312^21^221], when c_5 does not vanish;
[$2312^21^22^2$], when c_5 vanishes.

[2312^21^221]. The defining equation of this case can be reduced to
$$c_5 = 1.$$
This gives a quintuple algebra which may be called (ag_5), its multiplication table being *

(ag_5)	i	j	k	l	m
i	j	k	0	0	l
j	k	0	0	0	0
k	0	0	0	0	0
l	0	0	0	0	k
m	l	0	0	k	$j+k$

[$2312^21^22^2$]. The defining equation of this case is
$$c_5 = 0.$$
This gives a quintuple algebra which may be called (ah_5), its multiplication table being †

---

*On substituting $i_1 = i + \frac{1}{2}j + m$, $m_1 = i + \frac{1}{2}j - m$, this algebra falls apart into two of the form (b_4). [C. S. P.]

† On substituting $i_1 = i + m$, $m_1 = i - m$, $j_1 = j + l$, $l_1 = j - l$, this algebra falls apart into two of the form (b_3). [C. S. P.]

(ah_5)	i	j	k	l	m
i	j	k	0	0	l
j	k	0	0	0	0
k	0	0	0	0	0
l	0	0	0	0	k
m	l	0	0	k	j

[2312^212]. The defining equation of this case is
$$d_{51} = -1,$$
which gives
$$c_{51} = -1, \quad ml = -k.$$
There are two cases:

[2312^2121], when c_{51} does not vanish;
[2312^212^2], when c_{51} vanishes.

[2312^2121]. The defining equation of this case can be reduced to
$$c_{51} = 1.$$
This gives a quintuple algebra which may be called (ai_5), its multiplication table being *

(ai_5)	i	j	k	l	m
i	j	k	0	0	l
j	k	0	0	0	0
k	0	0	0	0	0
l	0	0	0	0	k
m	$k-l$	0	0	$-k$	$j+ck$

---

* In relative form, $i = A:C - B:F + C:E + D:G + E:G$, $j = A:E + C:G$, $k = A:G$, $l = A:F - B:G$. $m = A:B + A:D + B:E + C:F + aD:G + F:G$. [C. S. P.]

[$2312^2 12^2$]. The defining equation of this case is

$$mi = -l.$$

There are two cases:

[$2312^2 12^2 1$], when c_5 does not vanish;
[$2312^2 12^2 3$], when c_5 vanishes.

[$2312^2 12^2 1$]. The defining equation of this case can be reduced to

$$c_5 = 1.$$

This gives a quintuple algebra which may be called (aj_5), its multiplication table being*

(aj_5)	i	j	k	l	m
i	j	k	0	0	l
j	k	0	0	0	0
k	0	0	0	0	0
l	0	0	0	0	k
m	$-l$	0	0	$-k$	$j+k$

[$2312^2 12^3$]. The defining equation of this case is

$$m^2 = j.$$

This gives a quintuple algebra which may be called (ak_5), its multiplication table being †

---

* In relative form, $i = A:C + C:E + E:G - B:F$, $j = A:E + C:G$, $k = A:G$, $l = A:F - B:G$, $m = A:B + B:E + C:F + F:G + A:D + D:G$. [C. S. P.]

† In relative form, $i = A:C + C:D + D:F - B:E$, $j = A:D + C:F$, $k = A:F$, $l = A:E - B:F$, $m = A:B + B:D + C:E + E:F$. [C. S. P.]

(ak_5)	i	j	k	l	m
i	j	k	0	0	l
j	k	0	0	0	0
k	0	0	0	0	0
l	0	0	0	0	k
m	$-l$	0	0	$-k$	j

[2312^3]. The defining equations of this case are
$$ml = lm = 0, \quad m^2 = c_5 k.$$
There are two cases:

[$2312^3 1$], when d_{51} is not unity;
[2312^4], when d_{51} is unity.

[$2312^3 1$]. The defining equation of this case is
$$d_{51} \neq 1,$$
which gives
$i[(1-d_{51})m - c_{51}j] = (1-d_{51})l - c_{51}k, \quad i[(1-d_{51})l - c_{51}k] = 0,$
$[(1-d_{51})l - c_{51}k]i = 0, \quad [(1-d_{51})m - c_{51}j]i = d_{51}[(1-d_{51})l - c_{51}k],$
$[(1-d_{51})l - c_{51}k][(1-d_{51})m - c_{51}j] = 0,$
$[(1-d_{51})m - c_{51}j][(1-d_{51})l - c_{51}k] = 0,$
$[(1-d_{51})m - c_{51}j]^2 = (1-d_{51})^2 c_5 k;$

so that the substitution of $(1-d_{51})m - c_{51}j$ for m, and of $(1-d_{51})l - c_{51}k$ for l, is the same as to make
$$c_{51} = 0.$$
There are now two cases:

[$2312^3 1^2$], when c_5 does not vanish;
[$2312^3 12$], when c_5 vanishes.

[$2312^3 1^2$]. The defining equation of this case can be reduced to
$$m^2 = k.$$

This gives a quintuple algebra which may be called (al_5), its multiplication table being*

(al_5)	i	j	k	l	m
i	j	k	0	0	l
j	k	0	0	0	0
k	0	0	0	0	0
l	0	0	0	0	0
m	dl	0	0	0	k

[2312³12]. The defining equation of this case is

$$m^2 = 0.$$

This gives a quintuple algebra which may be called (am_5), its multiplication table being

(am_5)	i	j	k	l	m
i	j	k	0	0	l
j	k	0	0	0	0
k	0	0	0	0	0
l	0	0	0	0	0
m	dl	0	0	0	0

*In relative form, $i = A:B + B:C + C:D + dE:F$, $j = A:C + B:D$, $k = A:D$, $l = A:F$, $m = A:E + B:F + E:D$. [C. S. P.]

[2312⁴]. The defining equation of this case is
$$d_{51} = 1.$$
There are two cases:
   [2312⁴1], when c_{51} does not vanish;
   [2312⁵], when c_{51} vanishes.

[2312⁴1]. The defining equation of this case is easily reduced to
$$c_{51} = 1.$$
There are two cases:
   [2312⁴1²], when c_5 does not vanish;
   [2312⁴12], when c_5 vanishes.

[2312⁴1²]. The defining equation of this case is easily reduced to
$$m^2 = k.$$
This gives a quintuple algebra which may be called (an_5), its multiplication table being*

(an_5)	i	j	k	l	m
i	j	k	0	0	l
j	k	0	0	0	0
k	0	0	0	0	0
l	0	0	0	0	0
m	$l+k$	0	0	0	k

[2312⁴12]. The defining equation of this case is
$$m^2 = 0.$$
This gives a quintuple algebra which may be called (ao_5), its multiplication table being †

---

* In relative form, $i = A:E + A:B + B:C + C:D + E:F$, $j = A:C + B:D + A:F$, $k = A:D$, $l = A:F$, $m = A:C + A:E + E:D$. [C. S. P.]

† In relative form, $i = A:B + B:C + C:D + E:F$, $j = A:C + B:D$, $k = A:D$, $l = A:F$, $m = A:C + A:E + B:F$. [C. S. P.]

(ao_5)	i	j	k	l	m
i	j	k	0	0	l
j	k	0	0	0	0
k	0	0	0	0	0
l	0	0	0	0	0
m	$l+k$	0	0	0	0

[2312^5]. The defining equation of this case is
$$mi = l.$$
There are two cases:
    [2312^52], when c_5 does not vanish;
    [2312^6], when c_5 vanishes.

[2312^51]. The defining equation of this case can be reduced to
$$m^2 = k.$$
This gives a quintuple algebra which may be called (ap_5), its multiplication table being *

(ap_5)	i	j	k	l	m
i	j	k	0	0	l
j	k	0	0	0	0
k	0	0	0	0	0
l	0	0	0	0	0
m	l	0	0	0	k

* In relative form, $i = A:B + B:C + C:D + E:F$, $j = A:C + B:D$, $k = A:D$, $l = A:F$, $m = A:E + B:F + E:D$. [C. S. P.]

[2312⁶]. The defining equation of this case is
$$m^2 = 0.$$
This gives a quintuple algebra which may be called (aq_5), its multiplication table being

(aq_5)	i	j	k	l	m
i	j	k	0	0	l
j	k	0	0	0	0
k	0	0	0	0	0
l	0	0	0	0	0
m	l	0	0	0	0

[232]. The defining equation of this case is
$$im = 0,*$$
which gives
$$0 = jm = km,$$
and it may be assumed that
$$li = 0.$$
This gives
$$lj = lk = 0 = il^2 = l^2i = ilm = iml = mli = im.$$
There are two cases:
$$[232^1], \text{ when } mi = l;$$
$$[232^2], \text{ when } mi = 0.$$

[232¹]. The defining equation of this case is
$$mi = l,$$
which gives
$$0 = mj = mk, \quad lm = c_{45}k + d_{45}l + e_{45}m,$$
$$lmi = l^2 = e_{45}l, \quad 0 = l^4 = e_{45}l^3 = e_{45} = l^2, \quad m^2 = c_5k + d_5l + e_5m,$$
$$m^2i = ml = e_5l, \quad 0 = m^4l = e_5m^3l = e_5 = ml; \quad 0 = lm^2 = d_{45}lm = d_{45}.$$

---

* What is meant is that every quantity not involving powers of i is nilfaciend with reference to i. Hence, $il = 0$, also. [C. S. P.]

There are two cases:

[2321²], when c_{45} does not vanish;
[23212], when c_{45} vanishes.

[2321²]. The defining equation of this case can be reduced to
$$lm = k,*$$
which gives
$$m^2 = c_5 k, \quad (m - c_5 l)^2 = 0,$$
so that the substitution of $m - c_5 l$ for m is the same as to make
$$c_5 = 0.$$
This gives a quintuple algebra which may be called (ar_5), of which the multiplication table is

(ar_5)	i	j	k	l	m
i	j	k	0	0	0
j	k	0	0	0	0
k	0	0	0	0	0
l	0	0	0	0	k
m	l	0	0	0	0

[23212]. The defining equation of this case is
$$lm = 0.$$
There are two cases:

[232121], when d_5 does not vanish;
[23212²], when d_5 vanishes.

[232121]. The defining equation of this case can be reduced to
$$d_5 = 1.$$
There are two cases:

[232121²], when c_5 does not vanish;
[2321212], when c_5 vanishes.

[232121²]. The defining equation of this case can be reduced to
$$c_5 = 1.$$

---

* But $0 = im = mim = lm$. Thus, this case disappears, and the algebra (ar_5) is incorrect. [C. S. P.]

This gives a quintuple algebra which can be called (as_5), its multiplication table being*

(as_5)	i	j	k	l	m
i	j	k	0	0	0
j	k	0	0	0	0
k	0	0	0	0	0
l	0	0	0	0	0
m	l	0	0	0	$k+l$

[2321212]. The defining equation of this case is
$$c_5 = 0.$$
This gives a quintuple algebra which may be called (at_5), its multiplication table being

(at_5)	i	j	k	l	m
i	j	k	0	0	0
j	k	0	0	0	0
k	0	0	0	0	0
l	0	0	0	0	0
m	l	0	0	0	l

[23212^2]. The defining equation of this case is
$$m^2 = c_5 k.$$
There are two cases:
        [$23212^2 1$], when c_5 does not vanish;
        [23212^3], when c_5 vanishes.

---

*In relative form, $i = A:B + B:C + C:D + E:F$, $j = A:C + B:D$, $k = A:D$, $l = A:F$, $m = A:E + E:F + E:D$. Omitting the last term of m, we have (at_5). [C. S. P.]

[23212^21]. The defining equation of this case can be reduced to
$$m^2 = k.$$
This gives a quintuple algebra which may be called (au_5), its multiplication table being*

(au_5)	i	j	k	l	m
i	j	k	0	0	0
j	k	0	0	0	0
k	0	0	0	0	0
l	0	0	0	0	0
m	l	0	0	0	k

[23212^3]. The defining equation of this case is
$$m^2 = 0.$$
This gives a quintuple algebra which may be called (av_5), its multiplication table being

(av_5)	i	j	k	l	m
i	j	k	0	0	0
j	k	0	0	0	0
k	0	0	0	0	0
l	0	0	0	0	0
m	l	0	0	0	0

*In relative form, $i = A:B + B:C + C:D$, $j = A:C + B:D$, $k = A:D$, $l = E:D$, $m = E:C + A:F + F:D$. The omission of the last two terms of m gives (av_5). [C. S. P.]

[232']. The defining equation of this case is
$$mi = 0,$$
which gives
$$0 = mj = mk = lmi = m^2 i,$$
and there is no pure algebra in this case.

[24]. The defining equation of this case is
$$i^3 = 0,$$
and by § 59,
$$i^2 = j, \quad ij = ji = j^2 = 0.$$

There are three cases:
[241], when $ik = l$, $il = m$;
[242], when $ik = l$, $il = im = 0$;
[243], when $ik = il = im = 0$.

[241]. The defining equations of this case are
$$ik = l, \quad il = m,$$
which give
$jk = m$, $im = jl = jm = 0$, $0 = iml = ml^2 = e_{54} ml^2 = e_{54}$, $jk = m$,
$il^2 = ml = b_{54} j$, $l^2 = b_{54} i + b_4 j + e_4 m$, $0 = l^3 = b_{54} m + e_4 ml = b_{54} = ml$,
$im^2 = 0$, $m^2 = b_5 j + e_5 m$, $0 = m^3 = e_5 m^2 = e_5$,
$imi = 0$, $mi = b_{51} j + e_{51} m$, $mj = e_{51} mi$, $mi^3 = 0 = e_{51}$,
$ili = mi = b_{51} j$, $li = b_{51} i + b_{41} j + e_{41} m$,
$lil = lm = b_{51} m$, $0 = l^3 m = b_{51} = lm = mi = mil = m^2$, $(li)i = lj$,
$ik^2 = lk = a_3 j + c_3 l + d_3 m$, $ilk = mk = c_3 m$, $lik = l^2 = a_{31} m$,
$0 = mk^3 = c_3^2 m = c_3 = mk = k^2 m$,
$kj = ki^2 = a_{31} j + d_{31} li = a_{31}(1 + d_{31}) j + d_{31}^2 m$,
$kil = km = a_{31}(1 + d_{31}) m$, $0 = k^2 m = a_{31}(1 + d_{31}) km = a_{31}(1 + d_{31}) = km$,
$kj = d_{31}^2 m$, $0 = k^3 = a_3 l + b_3 m + d_3 lk = a_3 = b_3 + d_3^2 = b_3 kj + d_3 kl$,
$kl = a_{31} l + (b_{31} + d_3 d_{31}) m$, $0 = klk = a_{31} lk = d_3 a_{31} = lkl = a_{31} l^2 = a_{31} = l^2$,
$0 = d_3 a_{31} = b_{31} d_3 + d_3^2 d_{31} + b_3 d_{31}^2$,
*$0 = ki^3 + iki + i^2 k = (d_{31}^2 + d_{31} + 1) m$, $d_{31} = \sqrt[3]{1} = \mathbf{r}$,
$0 = k^2 i + kik + ik^2 = b_{31} + d_3 (1 + 2d_{31})$, $i(k + pi) = l + pj$, $i(l + pj) = m$,
$(k + pi) i = b_{31} j + d_{31} l + e_{31} m + pj = (b_{31} + p - pd_{31}) j + d_{31}(l + pj) + e_{31} m$,
$(l + pj) i = d_{31} m$,

---

* This line and the first equation of the next can be derived from $0 = (i + jk)^3$. [C. S. P.]

so that if p satisfies the equation
$$p(d_{31} - 1) = b_{31},$$
the substitution of $k + pi$ for k and of $l + pj$ for l is the same as to make
$$0 = b_{31} = d_3 = b_3.$$
There are four cases:

[241¹], when neither e_{31} nor e_3 vanishes;
[2412], when e_{31} does not vanish but e_3 vanishes;
[2413], when e_{31} vanishes and not e_3;
[2414], when e_{31} and e_3 both vanish.

[241¹]. The defining equations of this case can be reduced, without loss of generality, to
$$e_{31} = e_3 = 1.$$
We thus obtain a quintuple algebra which may be called (aw_5), its multiplication table being*

(aw_5)	i	j	k	l	m
i	j	0	l	m	0
j	0	0	m	0	0
k	$rl+m$	$r^2 m$	m	0	0
l	rm	0	0	0	0
m	0	0	0	0	0

[2412]. The defining equations of this case can be reduced to
$$e_{31} = 1, \quad e_3 = 0.$$

---

* In relative form $i = A:B + B:D + rC:E + rE:F + G:F$, $j = A:D + r^2 C:F$, $k = A:C + B:E + D:F + A:G + G:F$, $l = A:E + B:F$, $m = A:F$. To obtain (ax_5), omit the last term of k. To obtain (ay_5), omit, instead, the last term of i. To obtain (az_5), omit both these last terms. [C. S. P.]

We thus obtain a quintuple algebra which may be called (ax_5), its multiplication table being

(ax_5)	i	j	k	l	m
i	j	0	l	m	0
j	0	0	m	0	0
k	$rl+m$	r^2m	0	0	0
l	rm	0	0	0	0
m	0	0	0	0	0

[2413]. The defining equations of this case can be reduced to
$$e_{31} = 0, \quad e_3 = 1.$$
We thus obtain a quintuple algebra which may be called (ay_5), its multiplication table being

(ay_5)	i	j	k	l	m
i	j	0	l	m	0
j	0	0	m	0	0
k	rl	r^2m	m	0	0
l	rm	0	0	0	0
m	0	0	0	0	0

[2414]. The defining equations of this case are
$$e_{31} = e_3 = 0.$$
We thus obtain a quintuple algebra which may be called (az_5), its multiplication table being

(az_5)	i	j	k	l	m
i	j	0	l	m	0
j	0	0	m	0	0
k	rl	r^2m	0	0	0
l	rm	0	0	0	0
m	0	0	0	0	0

[242]. The defining equations of this case are

$$ik = l, \quad il = im = 0,$$

which give

$$0 = jk = jl = jm,$$
$$li = iki = a_{31}j + c_{31}l, \quad 0 = li^2 = c_{31} = li^3 = lj,$$
$$0 = a_{31}jk = lik = l^2 = ikl = a_{34} = c_{34},$$
$$ik^2 = lk = a_3j + c_3l, \quad 0 = ik^3 = lk^2 = c_3lk = c_3,$$
$$0 = imi = a_{51} = c_{51}, \quad mi^2 = mj = d_{51}li + e_{51}mi, \quad 0 = mji = e_{51},$$
$$mj = a_{31}d_{51}j, \quad 0 = m^2j = a_{31}d_{51} = mj,$$
$$ikm = lm = a_{35}j + c_{35}l, \quad 0 = lm^2 = c_{35},$$
$$0 = i^2k + iki + ki^2 = e_{31}d_{51} = 2a_{31} + a_{31}d_{31} + e_{31}b_{51},$$
$$kj = -a_{31}j, \quad 0 = k^2j = a_{31} = kj = li = e_{31}b_{51},$$
$$0 = imk = a_{53} = c_{53}, \quad 0 = mk^2 = e_{53},$$
$$mik = ml = a_3d_{51}j, \quad kik = kl = (a_3d_{31} + e_{31}b_{53})j + e_{31}d_{53}l, \quad e_{34} = 0,$$
$$0 = k^2l = e_{31}d_{53} = lki = e_{31}lm = e_{31}a_{35} = k^2m = e_{35}.$$

There are two cases:

[2421], when e_{31} does not vanish;
[242²], when c_{31} vanishes.

[2421]. The defining equation of this case can be reduced to

$$ki = m,$$

which, by the aid of the above equations, gives

$$0 = mi = kil = ml = kim = m^2, \quad a_3 j = ik^2 = k^2 i = lk = km, \quad 0 = lm,$$
$$b_{53} j = kik = kl = mk, \quad 0 = ik^2 + kik + k^2 i = 2a_3 + b_{53},$$
$$0 = k^2 = a_3 = b_{53} = kl = km = mk = ml;$$

and if p is determined by the equation

$$p^2 + (d_3 + e_3)p - b_3 = 0,$$

$k + pi$, $l + pj$, and $m + pj$ can be respectively substituted for k, l and m, which is the same thing as to make
$$b_3 = 0.$$
There are three cases:

[2421²], when neither d_3 nor e_3 vanishes;
[24212], when d_3 vanishes and not e_3;
[24213], when d_3 and e_3 both vanish.

[2421²]. The defining equation of this case can be reduced to

$$d_3 = 1.$$

This gives a quintuple algebra which may be called (ba_5), its multiplication table being*

(ba_5)	i	j	k	l	m
i	j	0	l	0	0
j	0	0	0	0	0
k	m	0	$l+em$	0	0
l	0	0	0	0	0
m	0	0	0	0	0

---

*In relative form, $i = A:B + B:C + A:E$, $j = A:C$, $k = D:B + E:F + D:G + eG:C + A:E$. $l = A:F$, $m = D:C$. By omitting the last term of k and putting $e = 1$ we get (bb_5), and by omitting the last two terms of k we get (bc_5). [C. S. P.]

[2421²]. The defining equation of this case can be reduced to

$$k^2 = m.$$

This gives a quintuple algebra which may be called (bb_5), its multiplication table being

(bb_5)	i	j	k	l	m
i	j	0	l	0	0
j	0	0	0	0	0
k	m	0	m	0	0
l	0	0	0	0	0
m	0	0	0	0	0

[24213]. The defining equation of this case is

$$k^2 = 0.$$

This gives a quintuple algebra which may be called (bc_5), its multiplication table being

(bc_5)	i	j	k	l	m
i	j	0	l	0	0
j	0	0	0	0	0
k	m	0	0	0	0
l	0	0	0	0	0
m	0	0	0	0	0

[242^3]. The defining equation of this case is
$$e_{31} = 0.$$

There are two cases:
[242^21], when e_3 does not vanish;
[242^3], when e_3 vanishes.

[242^21]. The defining equation of this case can be reduced to
$$k^2 = a_3 i + m,$$
which gives

$kik = kl = a_3 d_{31} j$, $ik^2 = lk = a_3 j$, $k^2 i = a_3 j + mi = d_{31}kl = a_3 d_{31}^2 j$,
$0 = k^2 i + ik + kik = a_3(d_{31}^2 + d_{31} + 1)$, $mi = a_3(d_{31}-1)j$, $ml = mik = 0$,
$0 = k^3 = a_3 l + mk = a_3 ki + km$,
$mk = -a_3 l$, $km = -a_3 b_{31} j - a_3 d_{31} l$, $lm = 0$.

There are two cases:
[242^21^2], when a_3 does not vanish;
[242^212], when a_3 vanishes.

[242^21^2]. The defining equation of this case can be reduced to
$$k^2 = i + m,$$
which gives

$d_{31} = \sqrt[3]{1} = \mathfrak{r}$, $lk = j$, $mk = -l$,
$ki = -km = b_{31} j + \mathfrak{r} l$, $mi = (\mathfrak{r}^2 - 1)j$, $m^2 = -\mathfrak{r} j$.

There are two cases:
[242^21^3], when b_{31} does not vanish;
[242^21^22], when b_{31} vanishes.

[242^21^3]. The defining equation of this case can be reduced to
$$ki = j + \mathfrak{r} l.$$

This gives a quintuple algebra which may be called (bd_5), its multiplication table being*

---

*In relative form, $i = A:D + D:F + B:E + C:F$, $j = A:F$, $k = \mathfrak{r}A:B + \mathfrak{r}B:C + D:E - \frac{1}{\mathfrak{r}}D:F + E:F$, $i = A:E - \frac{1}{\mathfrak{r}}A:F + B:F$, $m = \mathfrak{r}^2 A:C - A:D - B:E - C:F$. [C.S.P.]

l

(bd_5)

	i	j	k	l	m
i	j	0	l	0	0
j	0	0	0	0	0
k	$j+\mathfrak{r}l$	0	$i+m$	$\mathfrak{r}j$	$-j-\mathfrak{r}l$
l	0	0	j	0	0
m	$(\mathfrak{r}^2-1)j$	0	$-l$	0	$-\mathfrak{r}^2 j$

[$242^2 1^2 2$]. The defining equation of this case is
$$ki = \mathfrak{r}l.$$

This gives a quintuple algebra which may be called (be_5), its multiplication table being*

(be_5)

	i	j	k	l	m
i	j	0	l	0	0
j	0	0	0	0	0
k	$\mathfrak{r}l$	0	$i+m$	$\mathfrak{r}j$	$-\mathfrak{r}l$
l	0	0	j	0	0
m	$(\mathfrak{r}^2-1)j$	0	$-l$	0	$-\mathfrak{r}^2 j$

[$242^2 12$]. The defining equation of this case is
$$k^2 = m,$$
which gives
$$0 = kl = lk = km = mk = m^2 = k^2 i = mi.$$

There are two cases:

[$242^2 121$], when b_{31} does not vanish;
[$242^2 12^2$], when b_{31} vanishes.

---

* On adding to the expression for k in the last note the term $-A:C$, we have this algebra in relative form. [C. S. P.]

[$242^2 121$]. The defining equation of this case can be reduced to
$$ki = j + d_{31}l.$$
This gives a quintuple algebra which may be called (bf_5), its multiplication table being*

(bf_5)	i	j	k	l	m
i	j	0	l	0	0
j	0	0	0	0	0
k	$j+dl$	0	m	0	0
l	0	0	0	0	0
m	0	0	0	0	0

[$242^2 12^2$]. The defining equation of this case is
$$ki = d_{31}l.$$
This gives a quintuple algebra which may be called (bg_5), its multiplication table being †

(bg_5)	i	j	k	l	m
i	j	0	l	0	0
j	0	0	0	0	0
k	dl	0	m	0	0
l	0	0	0	0	0
m	0	0	0	0	0

---

* In relative form, $i = A:B + B:C + D:E$, $j = A:C$, $k = A:B + dA:D + B:E + B:F$. $l = A:E$, $m = A:E + A:F$. [C. S. P.]

† In relative form, $i = A:B + B:C + D:E$, $j = A:C$, $k = dA:D + B:E + B:F$, $l = A:E$, $m = A:F$. The algebra (ar_5) is what this becomes when $d = 0$. [C. S. P.]

[242^3]. The defining equation of this case is

$$e_3 = 0,$$

which gives *

---

* It is not easy to see how the author proves that $a_3 = 0$. But it can be proved thus. $0 = k^2 = (a_2 i + b_2 j + d_2 l)k = a_2 l + a_2 d_2 j$.

The algebras of the case [242^3] are those quintuple systems in which every product containing j or l as a factor vanishes, while every product which does not vanish is a linear function of j and l. Any multiplication table conforming to these conditions is self-consistent, but it is a matter of some trouble to exclude every case of a *mixed* algebra. An algebra of the class in question is separable, if all products are similar. But this case requires no special attention; and the only other is when two dissimilar expressions U and V can be found, such that both being linear functions of i, k and m, $UV = VU = 0$. It will be convenient to consider separately, first, the conditions under which $UV - VU = 0$, and, secondly, those under which $UV + VU = 0$. To bring the subjects under a familiar form, we may conceive of i, k, m as three vectors not coplanar, so that, writing

$$U = xi + yk + zm, \qquad V = x'i + y'k + z'm,$$

we have x, y, z, and x', y', z', the Cartesian coördinates of two points in space. [We might imagine the space to be of the hyperbolic kind, and take the coëfficients of j and l as coördinates of a point on the quadric surface at infinity. But this would not further the purpose with which we now introduce geometric conceptions.] But since we are to consider only such properties of U and V as belong equally to all their numerical multiples, we may assume that they always lie in any plane

$$Ax + By + Cz = 1,$$

not passing through the origin; and then x, y, z, and x', y', z', will be the homogeneous coördinates of the two points U and V in that plane. Let it be remembered that, although i, k, m are vectors, yet their multiplication does not at all follow the rule of quaternions, but that

$$i^2 = b_1 j + d_1 l, \qquad ik = b_{13} j + d_{13} l, \qquad im = b_{15} j + d_{15} l,$$
$$ki = b_{31} j + d_{31} l, \qquad k^2 = b_3 j + d_3 l, \qquad km = b_{35} j + d_{35} l,$$
$$mi = b_{51} j + d_{51} l, \qquad mk = b_{53} j + d_{53} l, \qquad m^2 = b_5 j + d_5 l.$$

The condition that $UV - VU = 0$ is expressed by the equations

$$(b_{13} - b_{31})(xy' - x'y) + (b_{15} - b_{51})(xz' - x'z) + (b_{35} - b_{53})(yz' - y'z) = 0,$$
$$(d_{13} - d_{31})(xy' - x'y) + (d_{15} - d_{51})(xz' - x'z) + (d_{35} - d_{53})(yz' - y'z) = 0.$$

The first equation evidently signifies that for every value of U, V must be on a straight line, that this line passes through U, and that it also passes through the point

$$P = (b_{35} - b_{53})i + (b_{51} - b_{15})k + (b_{13} - b_{31})m.$$

The second equation expresses that the line between U and V contains the point

$$Q = (d_{35} - d_{53})i + (d_{51} - d_{15})k + (d_{13} - d_{31})m.$$

The two equations together signify, therefore, that U and V may be any two points on the line between the fixed points P and Q. Linear transformations of j and l may shift P and Q to any other situations on the line joining them, but cannot turn the line nor bring the two points into coincidence.

The condition that $UV + VU = 0$ is expressed by the equations

$$2b_1 xx' + 2b_3 yy' + 2b_5 zz' + (b_{13} + b_{31})(xy' + x'y) + (b_{15} + b_{51})(xz' + x'z) + (b_{35} + b_{53})(yz' + y'z) = 0,$$
$$2d_1 xx' + 2d_3 yy' + 2d_5 zz' + (d_{13} + d_{31})(xy' + x'y) + (d_{15} + d_{51})(xz' + x'z) + (d_{35} + d_{53})(yz' + y'z) = 0.$$

The first of these evidently signifies that for any position of V the locus of U is a line; that U being fixed at any point on that line, V may be carried to any position on a line passing through its original position; and that further, if U is at one of the two points where its line cuts the conic

$$b_1 x^2 + b_3 y^2 + b_5 z^2 + (b_{13} + b_{31})xy + (b_{15} + b_{51})xz + (b_{35} + b_{53})yz = 0,$$

$$0 = k^2i = a_3j = a_3 = lk = ml = kl = m^3 = e_5 = d_5a_{35} = k^3m = c_{35},$$
$$0 = kmk = a_{35}l = a_{35} = lm.$$

---

then V may be at an infinitely neighboring point on the same conic, so that tangents to the conic from V cut the locus of U at their points of tangency. The second equation shows that the points U and V have the same relation to the conic

$$d_1x^2 + d_2y^2 + d_3z^2 + (d_{13}+d_{31})xy + (d_{15}+d_{31})xz + (d_{13}+d_{51})yz = 0.$$

These conics are the loci of points whose squares contain respectively no term in j and no term in l. Their four intersections represent expressions whose squares vanish. Hence, linear transformations of j and l will change these conics to any others of the sheaf passing through these four fixed points. The two equations together, then, signify that through the four fixed points, two conics can be drawn tangent at U and V to the line joining these last points.

Uniting the conditions of $UV - VU = 0$ and $UV + VU = 0$, they signify that U and V are on the line joining P and Q at those points at which this line is tangent to conics through the four fixed points whose squares vanish. But if the algebra is pure, it is impossible to find two such points; so that the line between P and Q must pass through one of the four fixed points. In other words, the necessary condition of the algebra being pure is that one and only one nilpotent expression in i. k. m. should be a linear function of P and Q.

The two points P and Q together with the two conics completely determine all the constants of the multiplication table. Let S and T be the points at which the two conics separately intersect the line between P and Q. A linear transformation of j will move P to the point $pP + (1-p)Q$ and will move S to the point $pS + (1-p)T$, and a linear transformation of l will move Q and T in a similar way. The points P and S may thus be brought into coincidence, and the point Q may be brought to the common point of intersection of the two conics with the line from P to Q. The geometrical figure determining the algebra is thus reduced to a first and a second conic and a straight line having one common intersection. This figure will have special varieties due to the coincidence of different intersections, etc.

There are six cases: [1], there is a line of quantities whose squares vanish and one quantity out of the line; [2], there are four dissimilar quantities whose squares vanish: [3], two of these four quantities coincide; [4], two pairs of the four quantities coincide; [5], three of the four quantities coincide; [6], all the quantities coincide.

We may, in every case, suppose the equation of the plane to be $x + y + z = 1$.

[1]. In this case, the line common to the two conics may be taken as $y = 0$, and the separate lines of the conics as $z = 0$ and $x = 0$, respectively. We may also assume $2P = x + y$ and $2Q = x + z$. We thus obtain the following multiplication table, where the rows and columns having j and l as their arguments are omitted:

	i	k	m
i	0	$3l$	$-j$
k	$-l$	0	$3j+l$
m	j	$l-j$	0

[2]. In this case, we may take k as the common intersection of the two conics and the line, i, m, and $i - k + m$ as the other intersections of the conics. We have $Q = k$, and we may write

$$P = S = pi + (1-p-q)k + qm, \qquad T = rP + (1-r)Q = rpi + (1-rp-rq)k + rqm.$$

We thus obtain the following multiplication table:

There are two cases:
[242^31], when d_5 does not vanish;
[242^4], when d_5 vanishes.

[242^31]. The defining equation of this case can be reduced to
$$d_5 = 1,$$
which gives
$$i(k + b_5 i + pm) = l + b_5 j = m^2;$$
and if
$$p^2 = -d_3 - b_5 - b_5 d_{31} - b_5 p d_{51} - p d_{35} - p d_{53},$$

	i	k	m
i	0	$q(q+1)j + rq(rq-1)l$	$[-2-p(p-3)+q(q+1)]j$ $+[-2-rp(rp-1)+rq(rq-1)]l$
k	$q(q-3)j + rq(rq-1)l$	0	$-p(p-3)j - rp(rp-1)l$
m	$[2-p(p+1)+q(q-3)]j +$ $[2-rp(rp-1)+rq(rq-1)]l$	$-p(p+1)j - rp(rp-1)l$	0

[3]. Let k be the double point common to the two conics, and let i and m be their other intersections. Then all expressions of the form $ku + uk$ are similar. The line between P and Q cannot pass through k, because in that case all products would be similar. We may therefore assume that it passes through i. Then, we have $Q = i$. we may assume $S = P = i - k + m$, and we may write $T = rP + (1-r)Q = i - rk + rm$. The equation of the common tangent to the conics at k may be written $hx + (1-h)z = 0$. Then the equations of the two conics are
$$hxy + xz + (1-h)yz = 0,$$
$$hxy + (h + r - hr)xz + (1-h)yz = 0.$$

We thus obtain the following multiplication table:

	i	k	m
i	0	$(h+1)j + (h+r)l$	$2j + [h(1-r) + 2r]l$
k	$(h-1)j + (h-r)l$	0	$(2-h)(j+l)$
m	$h(1-r)l$	$-h(j+l)$	0

[4]. In this case we may take i and m as the two points of contact of the conics, k as P, and $i - k + m$ as T. Then writing the equations of the two tangents

$$gy + z = 0. \qquad x + hy = 0.$$

the two conics become
$$gxy + xz + hyz = 0.$$
$$(g + h - 1)y^2 + gxy + xz + hyz = 0.$$

and the multiplication table is as follows:

PEIRCE: *Linear Associative Algebra.* 97

the substitution of $k + b_5 i + pm$ for k and $l + b_5 j$ for l is the same as to make

$$b_5 = d_3 = 0.$$

This gives

$$m^2 = l, \quad k^2 = b_3 j.$$

There are two cases:

[$242^3 1^2$], when b_3 does not vanish;
[$242^3 12$], when b_3 vanishes.

[$242^3 1^2$]. The defining equation of this case can be reduced to

$$k^2 = j.$$

	i	k	m
i	$(g+h-1)l$	$gj+(g+1)l$	$2l$
k	$gj+(g-1)l$	0	$hj+(h+1)l$
m	$2j$	$hj+(h-1)l$	0

[5]. In this case, we may take k as the point of osculation of the conics and i as their point of intersection. The line between P and Q must either, [51], pass through k, or, [52], pass through i.

[51]. We may, without loss of generality, take

$$P = k, \quad Q = m,$$

and the equations of the two conics are

$$z^2 + rxz = 0, \quad rxy + 2qxz + 2yz = 0.$$

Then, the multiplication table is as follows:

	i	k	m
i	0	0	ql
k	rl	0	l
m	$rj+ql$	l	j

[52]. We have $Q = i$, we may take $T = m$, and we may assume $P = 2i - m$ and $b_{13} + b_{31} = 1$. Then, we may write the equations of the two conics,

$$2z^2 + xy + xz + ryz = 0,$$
$$-rxy + (2-r)xz + r^2 yz = 0.$$

We thus obtain the following multiplication table:

This gives a quintuple algebra which can be called (bh_5), its multiplication table being*

(bh_5)	i	j	k	l	m
i	j	0	l	0	0
j	0	0	0	0	0
k	$aj+bl$	0	j	0	$cj+dl$
l	0	0	0	0	0
m	$a'j+b'l$	0	$c'j+d'l$	0	l

	i	k	m
i	0	$-rl$	$j-(r-2)l$
k	$2j-rl$	0	$(r-2)j+(r^2+1)l$
m	$j-(r-2)l$	$(r-2)j+(r^2-1)l$	$2j$

[6]. The conics have but one point in common. This may be taken at k. We have $Q=k$, we may take $T=i$ and $2P=2S=i+k$. We may also take $b_1=-1$. Then the equations of the two conics may be written
$$-x^2 + pz^2 + 2xy + 4qxz + 2ryz = 0,$$
$$(4+pr^2)z^2 + 2xy + 4(q+r)xz + 2ryz = 0.$$

We thus find this multiplication table:

	i	k	m
i	$-j$	$j+l$	$(2q-1)j+2(q+r-p)l$
k	$j+l$	0	$(r+1)j+rl$
m	$(2q+1)+2(q+r+p)l$	$(r-1)j+rl$	$pj+(4+pr^2)l$

If this analysis is correct, only three indeterminate coefficients are required for the multiplication tables of this class of algebras. [C. S. P.]

*See last note. I do not give relative forms for this class of algebras, owing to the extreme ease with which they may be found. [C. S. P.]

[242^312]. The defining equation of this case is

$$k^2 = 0.$$

There are two cases:

[242^3121], when b_{31} does not vanish;
[242^312^2], when b_{31} vanishes.

[242^3121]. The defining equation of this case can be reduced to

$$b_{31} = 1.$$

This gives a quintuple algebra which may be called (bi_5), its multiplication table being

(bi_5)	i	j	k	l	m
i	j	0	l	0	0
j	0	0	0	0	0
k	$j+al$	0	0	0	$bj+cl$
l	0	0	0	0	0
m	$a'j+b'l$	0	$cj+d'l$	0	l

[242^312^2]. The defining equation of this case is

$$ki = d_{31}l.$$

There are two cases:

[242^312^21], when b_{51} does not vanish;
[242^312^3], when b_{51} vanishes.

[242^312^21]. The defining equation of this case can be reduced to

$$b_{51} = 1.$$

This gives a quintuple algebra which may be called (bj_5), its multiplication table being

(bj_5)

	i	j	k	l	m
i	j	0	l	0	0
j	0	0	0	0	0
k	al	0	0	0	$bj+cl$
l	0	0	0	0	0
m	$j+a'l$	0	$b'j+il$	0	l

[$242^2 12^3$]. The defining equation of this case is

$$mi = d_{51} l;$$

which can always, in the case of a pure algebra, be reduced to

$$mi = l.$$

This gives a quintuple algebra which may be called (bk_5), its multiplication table being

(bk_5)

	i	j	k	l	m
i	j	0	l	0	0
j	0	0	0	0	0
k	al	0	0	0	$bj+cl$
l	0	0	0	0	0
m	l	0	$a'j+b'l$	0	l

[242^4]. The defining equation of this case is

$$m^2 = b_5 j,$$

and it can be reduced to [242^31] unless
$$d_{51} = d_3 = 0, \quad l^2 = b_3 j, \quad d_{31} = -1, \quad d_{35} = -d_{53};$$
whence it may be assumed that
$$mi = j;$$
and since
$$(k + bi)^2 = 0,$$
when
$$p^2 + p b_{31} + b_3 = 0,$$
it may also be assumed that
$$l^2 = 0.$$
There are two cases:

[242^41], when b_{31} does not vanish;
[242^5], when b_{31} vanishes.

[242^41]. The defining equation of this case can be reduced to
$$b_{31} = 1.$$

This gives a quintuple algebra which may be called (bl_5), its multiplication table being

(bl_5)	i	j	k	l	m
i	j	0	l	0	0
j	0	0	0	0	0
k	$j-l$	0	0	0	$aj+bl$
l	0	0	0	0	0
m	j	0	$aj+bl$	0	cj

[242^5]. The defining equation of this case is
$$ki = -l.$$
There are two cases:

[242^51], when b_{35} does not vanish;
[242^6], when b_{35} vanishes.

[242⁵1]. The defining equation of this case can be reduced to

$$b_{35} = 1.$$

This gives a quintuple algebra which may be called (bm_5), its multiplication table being*

(bm_5)	i	j	k	l	m
i	j	0	l	0	0
j	0	0	0	0	0
k	$-l$	0	0	0	$j+al$
l	0	0	0	0	0
m	j	0	$bj-al$	0	cj

[242⁶]. The defining equation of this case is

$$b_{35} = 0.$$

There are two cases:

    [242⁶1], when b_{53} does not vanish;
    [242⁷], when b_{53} vanishes.

[242⁶1]. The defining equation of this case can be reduced to

$$b_{53} = 1.$$

This gives a quintuple algebra which may be called (bn_5), its multiplication table being †

---

*This algebra is mixed. Namely, if $b \neq 1$, it separates on substituting $i_1 = (1-b)i+k$, $k_1 = (1-b)i + [a(1-b)+1]k - (1-b)^2 m$; but if $b=1$, it separates on substituting $i_1 = ai - (a^2 + a + c)k + m$, $k_1 = ai + qk + m$. [C. S. P.]

† Substitute $i_1 = i - k$, $k_1 = ak + m$, and the algebra separates. [C. S. P.]

(bn_5)	i	j	k	l	m
i	j	0	l	0	0
j	0	0	0	0	0
k	$-l$	0	0	0	al
l	0	0	0	0	0
m	j	0	$j-al$	0	cj

[242⁷]. The defining equation of this case is

$$b_{53} = 0.$$

This gives a quintuple algebra which may be called (bo_5), its multiplication table being*

(bo_5)	i	j	k	l	m
i	j	0	l	0	0
j	0	0	0	0	0
k	$-l$	0	0	0	al
l	0	0	0	0	0
m	j	0	$-al$	0	cj

[243]. The defining equations of this case are

$$0 = ik = il = im,$$

which give

$$0 = jk = jl = jm.$$

---

*Substitute for m, $ai + m$, and the algebra separates. [C. S. P.]

There are two cases:

[2431], when $ki = l$, $li = m$, $mi = 0$;
[2432], when $ki = l$, $li = mi = 0$.

[2431]. The defining equations of this case are

$$ki = l, \quad li = m, \quad mi = 0,$$

which give

$$kj = m, \quad lj = mj = 0 = lk = mk = l^2 = lm = ml = m^2,$$
$$0 = ik^2 = ikl = ikm = a_3 = a_{31} = a_{35},$$
$$k^2i = kl = c_3l + d_3m,$$
$$kli = km = c_3m, \quad 0 = k^3m = c_3 = km,$$
$$0 = k^3 = b_3m + d_3^2m = b_3 + d_3^2.$$

There are two cases:

[2431²], when c_3 does not vanish;
[24312], when c_3 vanishes.

[2431²]. The defining equation of this case can be reduced to

$$c_3 = 1.$$

This gives a quintuple algebra which may be called (b_5), its multiplication table being[*]

---

[*] The structure of this algebra may be exhibited by putting $k_1 = i + a^{-1}j - a^{-1}k$, $l_1 = j - a^{-1}l$, $m_1 = -a^{-1}m$, when the multiplication table becomes

	i	j	k	l	m
i	j	0	j	0	0
j	0	0	0	0	0
k	l	m	0	0	0
l	m	0	m	0	0
m	0	0	0	0	0

In relative form, $i = B:C + C:D$, $j = B:D$, $k = A:B + C:D$, $l = A:C$, $m = A:D$. [C. S. P.]

(bp_5)	i	j	k	l	m
i	j	0	0	0	0
j	0	0	0	0	0
k	l	m	$-a^2j+al+m$	am	0
l	m	0	0	0	0
m	0	0	0	0	0

[24312]. The defining equation of this case is

$$e_3 = 0.$$

This gives a quintuple algebra which may be called (bq_5), its multiplication table being*

(bq_5)	i	j	k	l	m
i	j	0	0	0	0
j	0	0	0	0	0
k	l	m	$-a^2j+al$	am	0
l	m	0	0	0	0
m	0	0	0	0	0

[2432]. The defining equations of this case are

$$ki = l, \quad li = mi = 0,$$

---

* On substituting $k_1 = i - a^{-1}k$, $l_1 = j - a^{-1}l$, $m_1 = a^{-1}m$, this algebra reduces to (bp_5), in the form given in the last note. [C. S. P.]

which give
$$0 = kj = lj = mj = lk = l^2 = lm = ik^2 = a_3,$$
$$kl = c_3 l, \quad 0 = k^2 l = c_3 = kl,$$
$$0 = ikm = a_{35} = kmi = c_{35} = k^2 m = e_{35} = imk = a_{53},$$
$$ml = c_{53} l, \quad 0 = m^2 l = c_{53} = ml = mk^2 = e_{53};$$

and it may be assumed that
$$k^2 = m,$$
which gives
$$0 = k^3 = km = mk = m^2.$$

There is then a quintuple algebra which may be called (br_5), its multiplication being*

(br_5)	i	j	k	l	m
i	j	0	0	0	0
j	0	0	0	0	0
k	l	0	m	0	0
l	0	0	0	0	0
m	0	0	0	0	0

[25]. The defining equations of this case are
$$0 = i^2 = j^2 = k^3 = l^2 = m^2 = ij + ji = ik + ki = il + li = im + mi,$$
$$0 = jk + kj = jl + lj = jm + mj = kl + lk = km + mk = lm + ml;$$
and it may be assumed that
$$ij = k = -ji, \quad il = m = -li;$$

---

*In relative form, $i = D:E + E:F$, $j = D:F$, $k = A:B + B:F + C:E$, $l = C:F$, $m = A:F$. [C. S. P.]

which gives

$$0 = ik = ki = jk = kj = im = mi = km = mk = lm = ml,$$
$$ijk = kl = b_{24}k + d_{24}m = -ilj = -mj = jm,$$
$$0 = j^2m = d_{24}jm = d_{24} = kl^2 = b_{24}kl = b_{24} = kl = lk = jm = mj,$$
$$0 = j^2l = a_{24}k = a_{24},$$
$$i(c_{24}j + e_{24}l) = c_{24}k + e_{24}m,$$
$$j(c_{24}j + e_{24}k) = e_{24}(c_{24}k + e_{24}m),$$
$$l(c_{24}j + e_{24}l) = -c_{24}(c_{24}k + e_{24}m);$$

so that it is easy to see that there is no pure algebra in this case.

SEXTUPLE ALGEBRA.

There are two cases:

[1], when there is an idempotent basis;
[2], when the algebra is nilpotent.

[1]. The defining equation of this case is

$$i^2 = i.$$

There are 19 cases:

[1²], when all the other units but i are in the first group;
[12], when j, k, l, m are in the first and n in the second group;
[13], when j, k and l are in the first and m and n in the second group;
[14], when j, k and l are in the first, m in the second and n in the third group;
[15], when j and k are in the first and l, m and n in the second group;
[16], when j and k are in the first, l and m in the second and n in the third group;
[17], when j and k are in the first, l in the second, m in the third, and n in the fourth group;
[18], when j is in the first, and k, l, m and n in the second group;
[19], when j is in the first, k, l and m in the second, and n in the third group;
[10′], when j is in the first, k and l in the second, and m and n in the third group;
[11′], when j is in the first, k and l in the second, m in the third and n in the fourth group;
[12′], when j is in the first, k in the second, l in the third and m and n in the fourth group; .

[13′], when j, k, l, m and n are in the second group;
[14′], when j, k, l and m are in the second and m in the third group;
[15′], when j, k and l are in the second and m and n are in the third group;
[16′], when j, k and l are in the second, m in the third, and n in the fourth group;
[17′], when j and k are in the second, l and m in the third, and n in the fourth group;
[18′], when j and k are in the second, l in the third, and m and n in the fourth group;
[19′], when j is in the second, k in the third, and l, m and n in the fourth group.

[1^2]; The defining equations of this case are

$$ij = ji = j, \quad ik = ki = k, \quad il = li = l, \quad im = mi = m, \quad in = ni = n,$$

and the 54 algebras of this case deduced from (q_5) to (br_5) may be called (a_6) to (bb_6).*

[12]. The defining equations of this case are

$$ij = ji = j, \quad ik = ki = k, \quad il = li = l, \quad im = mi = m, \quad in = n, \quad ni = 0,$$

which give

$$0 = jn = nj = kn = nk = ln = nl = mn = nm = n^2,$$

so that there is no pure algebra in this case.

[13]. The defining equations of this case are

$$ij = ji = j, \quad ik = ki = k, \quad il = li = l, \quad im = m, \quad in = n, \quad mi = ni = 0.$$

There are four cases, which correspond to relations between the units of the first group similar to those of the quadruple algebras (a_4), (b_4), (c_4) or (d_4).

[131]. The defining equations of this case are

$$j^2 = k, \quad jk = kj = l, \quad jl = k^2 = kl = lj = lk = l^2 = 0;$$

and, in the result, we obtain

$$jm = n, \quad jn = km = kn = lm = ln = 0.$$

---

* The multiplication tables of these algebras, formed from the nilpotent quintuple algebras, in the same manner in which the first class of quintuple algebras are formed from the nilpotent quadruple algebras, have been omitted. [C. S. P.]

This gives a sextuple algebra which may be called (bc_6), of which the multiplication table is*

(bc_6)	i	j	k	l	m	n
i	i	j	k	l	m	n
j	j	k	l	0	n	0
k	k	l	0	0	0	0
l	l	0	0	0	0	0
m	0	0	0	0	0	0
n	0	0	0	0	0	0

[132]. The defining equations of this case are

$$j^2 = k = l^2, \quad lj = ak, \quad jk = jl = kj = k^2 = kl = lk = 0,$$

which give

$$km = kn = 0.$$

There are two cases:

[132_1], when e_{24} does not vanish;
[132^2], when e_{24} vanishes.

[132_1]. The defining equation of this case can be reduced to

$$jn = m,$$

which gives

$$0 = jm = lm.$$

This gives a sextuple algebra which may be called (bd_6), of which the multiplication table is †

---

* In relative form, $i = A:A + B:B + C:C + D:D$, $j = A:B + B:C + C:D$, $k = A:C + B:D$, $l = A:D$, $m = B:E$, $n = A:E$. [C. S. P.]

† This algebra is distinguishable into two, in the same manner as (c_5). Namely, if $a = \pm 2$, on substituting $l_1 = l \pm j$, we have $l^2 = 0$, $jl = k$, $lj = -k$, and the multiplication table is otherwise unchanged. Otherwise, on substituting $j_1 = l + cj$, $l_1 = k + c^{-1}j$, where $2c = -a \pm \sqrt{a^2 - 4}$, we have $j^2 = l^2 = 0$, $jl = (1 - c^2)k$, $lj = (1 - c^{-2})k$, $jn = (b + c)k$, $ln = (b + c^{-1})k$, and otherwise the multiplication table is unchanged. The following is a relative form for the first variety: $i = A:A + B:B + C:C + D:D$, $j = A:B + B:D + C:D$, $k = A:D$, $l = A:C - B:D$, $m = A:E$, $n = B:E + bC:E$. For the second variety, $i = A:A + B:B + C:C + D:D$, $j = A:B + (1-c^2)C:D$, $k = A:D$, $l = A:C + (1-c^{-2})B:D$, $m = A:E$, $n = (b+c)B:E + (b+c^{-1})C:E$. [C. S. P.]

(bd_6)	i	j	k	l	m	n
i	i	j	k	l	m	n
j	j	k	0	0	0	m
k	k	0	0	0	0	0
l	l	ak	0	k	0	bm
m	0	0	0	0	0	0
n	0	0	0	0	0	0

[132^2]. The defining equation of this case is

$$jn = 0,$$

and there is no pure algebra in this case.

[13^2]. The defining equations of this case are

$$j^2 = k, \quad lj = k, \quad jk = jl = kj = k^2 = kl = lk = l^2 = 0,$$

which give

$$km = kn = 0.$$

There is a sextuple algebra in this case which may be called (be_6), of which the multiplication table is *

---

*This algebra may be a little simplified by substituting $j-l$ for j. In relative form, $i = A:A + B:B + C:C + D:D$, $j = A:D + B:C$, $k = A:C$, $l = A:B$, $m = A:E$, $n = bB:E + aD:E$.
[C. S. P.]

(be_6)	i	j	k	l	m	n
i	i	j	k	l	m	n
j	j	k	0	0	0	am
k	k	0	0	0	0	0
l	l	k	0	0	0	bm
m	0	0	0	0	0	0
n	0	0	0	0	0	0

[134]. The defining equations of this case are

$$jk = -kj = l, \quad j^2 = k^2 = jk = kj = kl = lk = l^2 = 0.$$

There is a sextuple algebra in this case which may be called (bf_6), of which the multiplication table is *

(bf_6)	i	j	k	l	m	n
i	i	j	k	l	m	n
j	j	0	0	k	0	m
k	k	0	0	0	0	0
l	l	$-k$	0	0	0	am
m	0	0	0	0	0	0
n	0	0	0	0	0	0

* In relative form, $i = A:A + B:B + C:C + D:D$, $j = A:B - C:D$, $k = A:D$, $l = A:C + B:D$, $m = A:E$, $n = B:E + aC:E$. [C. S. P.]

[14]. The defining equations of this case are

$$ij = ji = j, \quad ik = ki = k, \quad il = li = l, \quad im = m, \quad ni = n, \quad mi = in = 0,$$

which give

$$0 = jm = jn = km = kn = lm = ln = mj = nj = mk = nk = ml = nl = m^2 = nm = n^2.$$

There are four cases defined as in [13].

[141]. The defining equations of this case are

$$j^2 = k, \quad jk = kj = l, \quad jl = k^2 = kl = lj = lk = l^2 = 0,$$

which give

$$mn = d_{50}l.$$

There is a sextuple algebra which may be called (bg_6), of which the multiplication table is*

(bg_6)	i	j	k	l	m	n
i	i	j	k	l	m	0
j	j	k	l	0	0	0
k	k	l	0	0	0	0
l	l	0	0	0	0	0
m	0	0	0	0	0	l
n	n	0	0	0	0	0

[142]. The defining equations of this case are the same as in [132], which give

$$mn = c_{50}k.$$

---

* In relative form, $i = A:A + B:B + C:C + D:D$, $j = A:B + B:C + C:D$, $k = A:C + B:D$, $l = A:D$, $m = A:E$, $n = E:D$. [C. S. P.]

There is a sextuple algebra which may be called (bh_6), of which the multiplication table is *

(bh_6)	i	j	k	l	m	n
i	i	j	k	l	m	0
j	j	k	0	0	0	0
k	k	0	0	0	0	0
l	l	ak	0	k	0	0
m	0	0	0	0	0	k
n	n	0	0	0	0	0

[143]. The defining equations of this case are the same as in [13²]. There is a sextuple algebra which may be called (bi_6), of which the multiplication table is †

(bi_6)	i	j	k	l	m	n
i	i	j	k	l	m	0
j	j	k	0	0	0	0
k	k	0	0	0	0	0
l	l	k	0	0	0	0
m	0	0	0	0	0	k
n	n	0	0	0	0	0

---

* This algebra has two varieties, analogous to those of (c_3). The first is, in relative form, $i = A:A + B:B + C:C + D:D$, $j = A:B + B:C + A:D$, $k = A:C$, $l = -A:B + D:C$, $m = A:E$, $n = E:C$. The second in relative form is the same except that $j = A:B + b^{-1}D:C$, $l = A:D - bB:C$. [C. S. P.]

† This algebra may be slightly simplified by putting $j-l$ for j. Then, in relative form, $i = A:A + B:B + C:C$, $j = B:C$, $k = A:C$, $l = A:B$, $m = A:D$, $n = D:C$. [C. S. P.]

[14²]. The defining equations of this case are the same as in [134]. There is a sextuple algebra which may be called (bj_6), of which the multiplication table is *

(bj_6)	i	j	k	l	m	n
i	i	j	k	l	m	0
j	j	0	0	k	0	0
k	k	0	0	0	0	0
l	l	$-k$	0	0	0	0
m	0	0	0	0	0	k
n	n	0	0	0	0	0

[15]. The defining equations of this case are

$$ij = ji = j, \quad ik = ki = k, \quad il = l, \quad im = m, \quad in = n, \quad li = mi = ni = 0,$$

which give

$$j^2 = k, \quad 0 = jk = kj = k^2 = lj = lk = l^2 = lm = ln = mj = mk = ml = m^2$$
$$= mn = nj = nk = nl = nm = n^2.$$

There is a sextuple algebra which may be called (bk_6), of which the multiplication table is †

---

* $i = A:A + B:B + C:C + D:D$. $j = A:B - C:D$, $k = A:D$, $l = A:C + B:D$, $m = A:E$. $n = E:D$. [C. S. P.]

† In relative form, $i = A:A + B:B + C:C$, $j = A:B + B:C$, $k = A:C$, $l = C:D$, $m = B:D$, $n = A:D$. [C. S. P.]

(bk_0)	i	j	k	l	m	n
i	i	j	k	l	m	n
j	j	k	0	m	n	0
k	k	0	0	n	0	0
l	0	0	0	0	0	0
m	0	0	0	0	0	0
n	0	0	0	0	0	0

[16]. The defining equations of this case are

$ij = ji = j$, $ik = ki = k$, $il = l$, $im = m$, $ni = n$, $li = mi = in = 0$,

which give

$j^2 = k$, $0 = jk = jn = kj = k^2 = kn = lj = lk = l^2 = lm = mj = mk = nl = m^2$
$= nj = nk = nl = nm = n^2$.

There is a sextuple algebra which may be called (bl_0), of which the multiplication table is *

(bl_0)	i	j	k	l	m	n
i	i	j	k	l	m	0
j	j	k	0	m	0	0
k	k	0	0	0	0	0
l	0	0	0	0	0	k
m	0	0	0	0	0	0
n	n	0	0	0	0	0

* In relative form, $i = A:A + B:B + C:C$, $j = A:B + B:C$, $k = A:C$, $l = B:D + A:E$, $m = A:D$, $n = E:C$. [C. S. P.]

116   PEIRCE: *Linear Associative Algebra.*

[17]. The defining equations of this case are
$$ij = ji = j, \quad ik = ki = k, \quad il = l, \quad mi = m, \quad li = im = in = ni = 0.$$
There is no pure algebra in this case.

[18]. The defining equations of this case are
$$ij = ji = j, \quad ik = k, \quad il = l, \quad im = m, \quad in = n, \quad ki = li = mi = ni = 0.$$
There is no pure algebra in this case.

[19]. The defining equations of this case are
$$ij = ji = j, \quad ik = k, \quad il = l, \quad im = m, \quad ni = n, \quad in = kl = li = ni = n.$$
There is no pure algebra in this case.

[10′]. The defining equations of this case are
$$ij = ji = j, \quad ik = k, \quad il = l, \quad mi = m, \quad ni = n, \quad im = in = ki = li = 0.$$
There is no pure algebra in this case.

[11′]. The defining equations of this case are
$$ij = ji = j, \quad ik = k, \quad il = l, \quad mi = m, \quad im = li = in = ni = 0.$$
There is no pure algebra in this case.

[12′]. The defining equations of this case are
$$ij = ji = j, \quad ik = k, \quad li = l, \quad il = im = in = ki = mi = ni = 0.$$
There is no pure algebra in this case.

[13′]. The defining equations of this case are
$$ij = j, \quad ik = k, \quad il = l, \quad im = m, \quad in = n, \quad ji = ki = li = mi = ni = 0.$$
There is no pure algebra in this case.

[14′]. The defining equations of this case are
$$ij = j, \quad ik = k, \quad il = l, \quad im = m, \quad ni = n, \quad ji = ki = li = mi = in = 0.$$
There is no pure algebra in this case.

[15′]. The defining equations of this case are
$$ij = j, \quad ik = k, \quad il = l, \quad mi = m, \quad ni = n, \quad im = in = ji = ki = li = 0.$$
There is no pure algebra in this case.

[16']. The defining equations of this case are
$$ij = j, \quad ik = k, \quad il = l, \quad mi = m, \quad im = in = jk = kl = li = ni = 0.$$
There is no pure algebra in this case.

[17']. The defining equations of this case are
$$ij = j, \quad ik = k, \quad li = l, \quad mi = m, \quad il = im = in = ji = ki = ni = 0.$$
There is no pure algebra in this case.

[18']. The defining equations of this case are
$$ij = j, \quad ik = k, \quad li = l, \quad ji = ki = il = im = in = mi = ni = 0.$$
There are six cases:

[18'1], when $m^2 = m$, $mn = n$, $nm = 0$,
[18'2], when $m^2 = m$, $mn = 0$, $nm = n$,
[18'3], when $m^2 = n$, $mn = nm = 0$, $n^2 = m$,
[18'4], when $m^2 = m$, $mn = nm = n^2 = 0$,
[18'5], when $m^2 = n$, $m^3 = 0$,
[18'6], when $m^2 = n^2 = 0$.

[18'1]. The defining equations of this case are
$$m^2 = m, \quad mn = n, \quad nm = 0.$$
There are two cases:

[18'1²], when $ml = 0$;
[18'12], when $ml = l$.

[18'1²]. The defining equation of this case is
$$ml = 0.$$
There is no pure algebra in this case.

[18'12]. The defining equation of this case is
$$ml = l.$$
There are two cases:

[18'121], when $jm = j$;
[18'12²], when $jm = 0$.

[18'121]. The defining equation of this case is
$$jm = j.$$
There is a sextuple algebra which may be called (bm_6), of which the multiplication table is *

---

* In relative form, $i = A:A$, $j = A:B$, $k = A:C$, $l = B:A$, $m = B:B$, $n = B:C$. [C. S. P.]

(bm_6)	i	j	k	l	m	n
i	i	j	k	0	0	0
j	0	0	0	i	j	k
k	0	0	0	0	0	0
l	l	m	n	0	0	0
m	0	0	0	l	m	n
n	0	0	0	0	0	0

[18'12²]. The defining equation of this case is
$$jm = 0.$$
There is no pure algebra in this case.

[18'2]. The defining equations of this case are
$$m^2 = m, \quad mn = 0, \quad nm = n.$$
There are two cases:
[18'21], when $ml = l$;
[18'2²], when $ml = 0$.

[18'21]. The defining equation of this case is
$$ml = l.$$
There are two cases:
[18'21²], when $jm = j$;
[18'212], when $jm = 0$.

[18'21²]. The defining equation of this case is
$$jm = j.$$
There is no pure algebra in this case.

[18'212]. The defining equation of this case is
$$jm = 0.$$
There is no pure algebra in this case.

[18′2²]. The defining equation of this case is
$$ml = 0.$$
There is no pure algebra in this case.

[18′3]. The defining equations of this case are
$$m^2 = m, \quad mn = nm = 0, \quad n^2 = n.$$
There is no pure algebra in this case.

[18′4]. The defining equations of this case are
$$m^2 = m, \quad mn = nm = n^2 = 0.$$
There are two cases:
[18′41], when $jm = j$;
[18′42], when $jm = 0$.

[18′41]. The defining equation of this case is
$$jm = j.$$
There is no pure algebra in this case.

[18′42]. The defining equation of this case is
$$jm = 0.$$
There is no pure algebra in this case.

[18′5]. The defining equations of this case are
$$m^2 = n, \quad m^3 = 0.$$
There is no pure algebra in this case.

[18′6]. The defining equations of this case are
$$m^2 = n^2 = 0.$$
There is no pure algebra in this case.

[19′]. The defining equations of this case are
$$ij = j, \quad ki = k, \quad ji = ik = il = im = in = li = mi = ni = 0.$$
There is no pure algebra in this case.

[2]. The algebras belonging to this case are not investigated, because it is evident from § 69 that they are rarely of use unless combined with an idempotent basis, so as to give septuple algebras.

Natural Classification.

There are many cases of these algebras which may obviously be combined into natural classes, but the consideration of this portion of the subject will be reserved to subsequent researches.

ADDENDA.

I.

On the Uses and Transformations of Linear Algebra.

BY BENJAMIN PEIRCE.

[*Presented to the American Academy of Arts and Sciences, May 11. 1875.*]

Some definite interpretation of a linear algebra would, at first sight, appear indispensable to its successful application. But on the contrary, it is a singular fact, and one quite consonant with the principles of sound logic, that its first and general use is mostly to be expected from its want of significance. The interpretation is a trammel to the use. Symbols are essential to comprehensive argument. The familiar proposition that all A is B, and all B is C, and therefore all A is C, is contracted in its domain by the substitution of significant words for the symbolic letters. The A, B, and C, are subject to no limitation for the purposes and validity of the proposition; they may represent not merely the actual, but also the ideal, the impossible as well as the possible. In Algebra, likewise, the letters are symbols which, passed through a machinery of argument in accordance with given laws, are developed into symbolic results under the name of formulas. When the formulas admit of intelligible interpretation, they are accessions to knowledge; but independently of their interpretation they are invaluable as symbolical expressions of thought. But the most noted instance is the symbol called the impossible or imaginary, known also as the square root of minus one, and which, from a shadow of meaning attached to it, may be more definitely distinguished as the symbol of *semi-inversion*. This symbol is restricted to a precise signification as the representative of perpendicularity in quaternions, and this wonderful algebra of space is intimately dependent upon the special use of the symbol for its symmetry, elegance, and power. The immortal author of quaternions has shown that there are other significations which may attach to the symbol in other cases. But the strongest use of the symbol is to be found in its magical power of doubling the actual universe, and placing by its side an ideal universe, its exact counterpart, with which it can be compared and contrasted, and, by means of curiously connecting fibres, form with it an organic whole, from which modern analysis has developed her surpassing geometry. The letters or units of the linear algebras, or to use the better term proposed by Mr. Charles S. Peirce, the *vids* of these algebras, are fitted to perform a similar function each in its peculiar way. This is their primitive and perhaps will always be their principal use. It does not exclude the possibility of some special modes of interpretation, but, on the contrary, a higher philosophy, which believes in the capacity of the material universe for all expressions of human thought, will find, in the utility of the vids, an indication of their probable reality of interpretation. Doctor Hermann Hankel's alternate numbers, with Professor Clifford's applications to determinants, are a curious and interesting example of the possible advantage to be obtained from the new algebras. Doctor Spottiswoode in his fine, generous, and complete analysis of my own treatise before the London Mathematical Society in November of 1872, has regarded these numbers as quite different from the algebras discussed in my treatise, because they are neither linear nor limited. But there is no difficulty in reducing them to a linear form, and, indeed, my algebra (e_3) is the simplest case of Hankel's alternate numbers; and in any other case, in which n is the number of the Hankel elements employed, the complete number of vids of the corresponding linear algebra is $2^n - 1$. The limited character of the algebras which I have investigated may be regarded as an accident of the mode of discussion. There is, however, a large number of unlimited algebras suggested by the investigations, and Hankel's numbers themselves would have been a natural generalization from the proposition of § 65 of my algebra.* Another class of unlimited algebras, which would readily occur from the inspection of those which are given, is that in which all the powers of a vid are adopted as independent vids, and the highest power may either be zero, or unity, or the vid itself, and the zero power of the fundamental vid, *i. e.* unity itself, may also be retained as a vid. But I desire to draw especial attention to that class, which is also unlimited, and for which, when it was laid before the mathematical society of London in January of 1870, Professor Clifford proposed the appropriate name of *quadrates*.

---

* This remark is not intended as a foundation for a claim upon the Hankel numbers, which were published in 1867, three years prior to the publication of my own treatise.—B. P. [They were given much earlier under the name of *clefs* by Cauchy, and (substantially) at a still earlier date by Grassmann. —C. S. P.]

Quadrates.

The best definition of quadrates is that proposed by Mr. Charles S. Peirce. If the letters A, B, C, etc., represent absolute quantities, differing in quality, the vids may represent the relations of these quantities, and may be written in the form

$$(A:A)\ (A:B)\ (A:C)\ \ldots\ (B:A)\ (B:B)\ \ldots\ (C:A),\ \text{etc.}$$

subject to the equations

$$(A:B)\ (B:C) = (A:C)$$
$$(A:B)\ (C:D) = 0.$$

In other words, every product vanishes, in which the second letter of the multiplier differs from the first letter of the multiplicand; and when these two letters are identical, both are omitted, and the product is the vid which is compounded of the remaining letters, which retain their relative position.

Mr. Peirce has shown by a simple logical argument that the quadrate is the legitimate form of a complete linear algebra, and that all the forms of the algebras given by me must be imperfect quadrates, and has confirmed this conclusion by actual investigation and reduction. His investigations do not however dispense with the analysis by which the independent forms have been deduced in my treatise, though they seem to throw much light upon their probable use.

Unity.

The sum of the vids $(A:A)$, $(B:B)$, $(C:C)$, etc., extended so as to include all the letters which represent absolute quantities in a given algebra, whether it be a complete or an incomplete quadrate, has the peculiar character of being idempotent, and of leaving any factor unchanged with which it is combined as multiplier or multiplicand. This is the distinguishing property of unity, so that this combination of the vids can be regarded as unity, and may be introduced as such and called the *vid of unity*. There is no other combination which possesses this property.

But any one of the vids $(A:A)$, $(B:B)$, etc., or the sum of any of these vids is idempotent. There are many other idempotent combinations, such as

$$(A:A) + x(A:B),\quad y(A:B) + (B:B),$$
$$\tfrac{1}{2}(A:A) + \tfrac{1}{2}(A:B) + \tfrac{1}{2}(B:A) + \tfrac{1}{2}(B:B),$$

which may deserve consideration in making transformations of an algebra preparatory to its application.

Inversion.

A vid which differs from unity, but of which the square is equal to unity, may be called a *vid of inversion*. For such a vid when applied to some other combination transforms it; but, whatever the transformation, a repetition of the application restores the combination to its primitive form. A very general form of a vid of inversion is

$$(A:A) \pm (B:B) \pm (C:C) \pm \text{etc.},$$

in which each doubtful sign corresponds to two cases, except that at least one of the signs must be negative. The negative of unity might also be regarded as a symbol of inversion, but cannot take the place of an independent vid. Besides the above vids of inversion, others may be formed by adding to either of them a vid consisting of two different letters, which correspond to two of the one-lettered vids of different signs; and this additional vid may have any numerical coefficient whatever. Thus

$$(A:A) + (B:B) - (C:C) + x(A:C) + y(B:C)$$

is a vid of inversion.

The new vid which Professor Clifford has introduced into his biquaternions is a vid of inversion.

Semi-Inversion.

A vid of which the square is a vid of inversion, is a *vid of semi-inversion*. A very general form of a vid of semi-inversion is

$$(A:A) \pm (B:B) \pm \mathsf{J}(C:C) \pm \text{etc.}$$

in which one or more of the terms $(A:A)$, $(B:B)$, etc., have J for a coefficient. The combination

$$(A:A) \pm \mathsf{J}(B:B) + x(A:B) + \text{etc.}$$

is also a vid of semi-inversion. With the exception of unity, all the vids of Hamilton's quaternions are vids of semi-inversion.

The Use of Commutative Algebras.

Commutative algebras are especially applicable to the integration of differential equations of the first degree with constant coefficients. If i, j, k, etc., are the vids of such an algebra, while x, y, z, etc., are independent variables, it is easy to show that a solution may have the form $F(xi + yj + zk + \text{etc.})$, in which F is an arbitrary function, and i, j, k, etc., are connected by some simple equation. This solution can be developed into the form

$$F(xi + yj + zk + \text{etc.}) = Mi + Nj + Pk + \text{etc.}$$

in which M, N, P, etc., will be functions of x, y, z, etc., and each of them is a solution of the given equation. Thus in the case of Laplace's equation for the potential of attracting masses, the vids must satisfy the equation

$$i^2 + j^2 + k^2 = 0.$$

The algebra (a_3) of which the multiplication table is

	i	j	k
i	i	j	k
j	j	k	0
k	k	0	0

may be used for this case. Combinations i_1, j_1, k_1 of these vids can be found which satisfy the equation

$$i_1^2 + j_1^2 + k_1^2 = 0,$$

and if the functional solution

$$F(xi_1 + yj_1 + zk_1)$$

is developed into the form of the original vids

$$Mi + Nj + Pk,$$

M, N, and P will be independent solutions, of such a kind that the surfaces for which N and P are constant will be perpendicular to that for which M is constant, which is of great importance in the problems of electricity.

The Use of Mixed Algebras.

It is quite important to know the various kinds of pure algebra in making a selection for special use, but mixed algebras can also be used with advantage in certain cases. Thus, in Professor Clifford's biquaternions, of which he has demonstrated the great value, other vids can be substituted for unity and his new vid, namely their half sum and half difference, and each of the original vids of the quaternions can be multiplied by these, giving us two sets of vids, each of which will constitute an independent quadruple algebra of the same form with quaternions. Thus if i, j, k, are the primitive quaternion vids and w the new vid, let

$$a_1 = \tfrac{1}{2}(1 + w). \qquad a_2 = \tfrac{1}{2}(1 - w).$$
$$i_1 = a_1 i. \qquad i_2 = a_2 i.$$
$$j_1 = a_1 j. \qquad j_2 = a_2 j.$$
$$k_1 = a_1 k. \qquad k_2 = a_2 k.$$

Then since

$$a_1^2 = a_1. \qquad a_2^2 = a_2.$$
$$i_1^2 = j_1^2 = k_1^2 = -a_1. \qquad i_2^2 = j_2^2 = k_2^2 = -a_1.$$
$$i_1 j_1 = k_1 = -j_1 i_1. \qquad i_2 j_2 = k_2 = -j_2 i_2.$$
$$j_1 k_1 = i_1 = -k_1 j_1. \qquad j_2 k_2 = i_2 = -k_2 j_2.$$
$$k_1 i_1 = j_1 = -i_1 k_1. \qquad k_2 i_2 = j_2 = -i_2 k_2.$$
$$a_1 a_2 = 0 = a_2 a_1.$$
$$M_1 N_2 = 0 = N_2 M_1.$$

in which M_1 denotes any combination of the vids of the first algebra, and N_2 any combination of those of the second algebra. It may perhaps be claimed that these algebras are not independent, because the sum of the vids a_1 and a_2 is absolute unity. This, however, should be regarded as a fact of interpretation which is not apparent in the defining equations of the algebras.

II.

On the Relative Forms of the Algebras.
By C. S. Peirce.

Given an associative algebra whose letters are i, j, k, l, etc., and whose multiplication table is

$$i^2 = a_{11} i + b_{11} j + c_{11} k + \text{etc.}*$$
$$ij = a_{12} i + b_{12} j + c_{12} k + \text{etc.}$$
$$ji = a_{21} i + b_{21} j + c_{21} k + \text{etc.,}$$
$$\text{etc., etc.}$$

I proceed to explain what I call the relative form of this algebra.

---

*I have used a_{11}, etc., in place of the a_1, etc., used by my father in his text.

Let us assume a number of new units, A, I, J, K, L, etc., one more in number than the letters of the algebra, and every one except the first, A, corresponding to a particular letter of the algebra. These new units are susceptible of being multiplied by numerical coefficients and of being added together;* but they cannot be multiplied together, and hence are called *non-relative* units.

Next, let us assume a number of operations each denoted by bracketing together two non-relative units separated by a colon. These operations, equal in number to the square of the number of non-relative units, may be arranged as follows:

$$(A:A) \quad (A:I) \quad (A:J) \quad (A:K), \text{ etc.}$$
$$(I:A) \quad (I:I) \quad (I:J) \quad (I:K), \text{ etc.}$$
$$(J:A) \quad (J:I) \quad (J:J) \quad (J:K), \text{ etc.}$$

Any one of these operations performed upon a polynomial in non-relative units, of which one term is a numerical multiple of the letter following the colon, gives the same multiple of the letter preceding the colon. Thus, $(I:J)(aI+bJ+cK)=bI$.† These operations are also taken to be susceptible of associative combination. Hence $(I:J)(J:K) = (I:K)$; for $(J:K)K = J$ and $(I:J)J = I$, so that $(I:J)(J:K)K = I$. And $(I:J)(K:L) = 0$; for $(K:L)L = K$ and $(I:J)K = (I:J)(0.J+K) = 0.I = 0$. We further assume the application of the distributive principle to these operations; so that, for example,

$$\{(I:J) + (K:J) + (K:L)\}(aJ + bL) = aJ + (a+b)K.$$

Finally, let us assume a number of complex operations denoted by i', j', k', l', etc., corresponding to the letters of the algebra and determined by its multiplication table in the following manner:

$$i' = (I:A) + a_{11}(I:I) + b_{11}(J:I) + c_{11}(K:I) + \text{etc.}$$
$$+ a_{12}(I:J) + b_{12}(J:J) + c_{12}(K:J) + \text{etc.}$$
$$+ a_{13}(I:K) + b_{13}(J:K) + c_{13}(K:K) + \text{etc.} + \text{etc.}$$
$$j' = (J:A) + a_{21}(I:I) + b_{21}(J:I) + c_{21}(K:I) + \text{etc.}$$
$$+ a_{22}(I:J) + b_{22}(J:J) + c_{22}(K:J) + \text{etc.}$$
$$+ a_{23}(I:K) + b_{23}(J:K) + c_{23}(K:K) + \text{etc.} + \text{etc.}$$
$$k' = \text{etc.}$$

---

* Any one of them multiplied by 0 gives 0.   † If $b=0$, of course the result is 0.

Any two operations are equal which, being performed on the same operand, invariably give the same result. The ultimate operands in this case are the non-relative units. But any operations compounded by addition or multiplication of the operations i', j', k', etc., if they give the same result when performed upon A, will give the same result when performed upon any one of the non-relative units. For suppose $i'j'A = k'l'A$. We have

$$i'j'A = i'J = a_{12}I + b_{12}J + c_{12}K + \text{etc.}$$
$$k'l'A = k'L = a_{34}I + b_{34}J + c_{34}K + \text{etc.}$$

so that $a_{12} = a_{34}$, $b_{12} = b_{34}$, $c_{12} = c_{34}$, etc., and in our original algebra $ij = kl$. Hence, multiplying both sides of the equation into any letter, say m, $ijm = klm$. But

$$ijm = i(a_{25}i + b_{25}j + c_{25}k + \text{etc.}) = (a_{11}a_{25} + a_{12}b_{25} + a_{13}c_{25} + \text{etc.})i$$
$$+ (b_{11}a_{25} + b_{12}b_{25} + b_{13}c_{25} + \text{etc.})j + \text{etc.}$$

But we have equally

$$i'j'm'A = (a_{11}a_{25} + a_{12}b_{25} + a_{13}c_{25} + \text{etc.})I + (b_{11}a_{25} + b_{12}b_{25} + b_{13}c_{25} + \text{etc.})J + \text{etc.}$$

So that $i'j'm'A = k'l'm'A$. Hence, $i'j'M = k'l'M$. It follows, then, that if $i'j'A = k'l'A$, then $i'j'$ into any non-relative unit equals $k'l'$ into the same unit, so that $i'j' = k'l'$. We thus see that whatever equality subsists between compounds of the accented letters i', j', k', etc., subsists between the same compounds of the corresponding unaccented letters i, j, k, so that the multiplication tables of the two algebras are the same.* Thus, what has been proved is that any associative algebra can be put into relative form, *i. e.* (see my *brochure* entitled *A brief Description of the Algebra of Relatives*) that every such algebra may be represented by a matrix.

Take, for example, the algebra (bd_5). It takes the relative form
$i = (I:A) + (J:I) + (L:K)$, $j = (J:A)$,
$k = (K:A) + (J:I) + \mathfrak{r}(L:I) + (I:K) + (M:K) + \mathfrak{r}(J:L) - (J:M) - \mathfrak{r}(L:M)$,
$l = (L:A) + (J:K)$, $m = (M:A) + (\mathfrak{r}^2 - 1)(J:I) - (L:K) - \mathfrak{r}^2(J:M)$.

---

* A brief proof of this theorem, perhaps essentially the same as the above, was published by me in the *Proceedings of the American Academy of Arts and Sciences*, for May 11, 1875.

This is the same as to say that the general expression $xi + yj + zk + ul + vm$ of this algebra has the same laws of multiplication as the matrix

$$\begin{matrix}
0, & 0, & 0, & 0, & 0, & 0, \\
x, & 0, & 0, & z, & 0, & 0, \\
y, & \begin{matrix}x+z\\+(\mathfrak{r}^2-1)v,\end{matrix} & 0, & u, & \mathfrak{r}z, & -z-\mathfrak{r}^2v, \\
z, & 0, & 0, & 0. & 0, & 0, \\
u, & \mathfrak{r}z, & 0, & x-v, & 0, & -\mathfrak{r}z \\
v, & 0, & 0, & z, & 0, & 0.
\end{matrix}$$

Of course, every algebra may be put into relative form in an infinity of ways; and simpler ways than that which the rule affords can often be found. Thus, for the above algebra, the form given in the foot-note is simpler, and so is the following:

$i = (B:A) + (C:B) + (F:D) + (C:E), \quad j = (C:A),$
$k = (D:A) + (E:D) + (C:B) + \mathfrak{r}(F:B) + \mathfrak{r}(C:F),$
$l = (F:A) + (C:D), \quad m = (E:A) + (\mathfrak{r}^2-1)(C:B) - (B:A) - (F:D) - (C:E).$

These different forms will suggest transformations of the algebra. Thus, the relative form in the foot-note to (bd_5) suggests putting

$i_1 = i + m, \quad j_1 = \mathfrak{r}^2 j, \quad k_1 = k + \mathfrak{r}^{-1}i + \mathfrak{r}^{-1}m, \quad l_1 = \mathfrak{r}l + j, \quad m_1 = -m,$

when we get the following multiplication table, where ρ is put for $\mathfrak{r}^{-1}$:

	i	j	k	l	m
i	0	0	0	0	j
j	0	0	0	0	0
k	0	0	i	j	l
l	0	0	ρj	0	0
m	$\rho^2 j$	0	ρl	0	j

Ordinary algebra with imaginaries, considered as a double algebra, is, in relative form,
$$1 = (X:X) + (Y:Y), \quad J = (X:Y) - (Y:X).$$
This shows how the operation J turns a vector through a right angle in the plane of X, Y. Quaternions in relative form is
$$1 = (W:W) + (X:X) + (Y:Y) + (Z:Z),$$
$$i = (X:W) - (W:X) + (Z:Y) - (Y:Z),$$
$$j = (Y:W) - (Z:X) - (W:Y) + (X:Z),$$
$$k = (Z:W) + (Y:X) - (X:Y) - (W:Z).$$
We see that we have here a reference to a space of four dimensions corresponding to X, Y, Z, W.

III.

On the Algebras in which Division is Unambiguous.

By C. S. PEIRCE.

1. In the *Linear Associative Algebra*, the coefficients are permitted to be imaginary. In this note they are restricted to being real. It is assumed that we have to deal with an algebra such that from $AB = AC$ we can infer that $A = 0$ or $B = C$. It is required to find what forms such an algebra may take.

2. If $AB = 0$, then either $A = 0$ or $B = 0$. For if not, $AC = A(B + C)$, although A does not vanish and C is unequal to $B + C$.

3. The reasoning of § 40 holds, although the coefficients are restricted to being real. It is true, then, that since there is no expression (in the algebra under consideration) whose square vanishes, there must be an expression, i, such that $i^2 = i$.

4. By § 41, it appears that for every expression in the algebra we have
$$iA = Ai = A.$$

5. By the reasoning of §53, it appears that for every expression A there is an equation of the form
$$\Sigma_m (a_m A^m) + bi = 0.$$
But i is virtually arithmetical unity, since $iA = Ai = A$; and this equation may be treated by the ordinary theory of equations. Suppose it has a real root, a; then it will be divisible by $(A - a)$, and calling the quotient B we shall have
$$(A - ai) B = 0.$$

But $A - ai$ is not zero, for A was supposed dissimilar to i. Hence a product of finites vanishes, which is impossible. Hence the equation cannot have a real root. But the whole equation can be resolved into quadratic factors, and some one of these must vanish. Let the irresoluble vanishing factor be

Then
$$(A - s)^2 + t^2 = 0.$$
$$\left(\frac{A-s}{t}\right)^2 = -1,$$

or, every expression, upon subtraction of a real number (i. e. a real multiple of i), can be converted, in one way only, into a quantity whose square is a negative number. We may express this by saying that every quantity consists of a scalar and a vector part. A quantity whose square is a negative number we here call a *vector*.

6. Our next step is to show that the vector part of the product of two vectors is linearly independent of these vectors and of unity. That is, i and j being any two vectors,* if
$$ij = s + v$$
where s is a scalar and v a vector, we cannot determine three real scalars a, b, c, such that
$$v = a + bi + cj.$$
This is proved, if we prove that no scalar subtracted from ij leaves a remainder $bi + cj$. If this be true when i and j are any unit vectors whatever, it is true when these are multiplied by real scalars, and so is true of every pair of vectors. We will, then, suppose i and j to be unit vectors. Now,
$$ij^2 = -i.$$
If therefore we had
$$ij = a + bi + cj,$$
we should have
$$-i = ij^2 = aj + bij - c = ab - c + b^2i + (a + bc)j;$$
whence, i and j being dissimilar,
$$-i = b^2i, \qquad b^2 = -1,$$
and b could not be real.

---

* The idempotent basis having been shown to be arithmetical unity, we are free to use the letter i to denote another unit.

7. Our next step is to show that, i and j being any two vectors, and
$$ij = s + v,$$
s being a scalar and v a vector, we have
$$ji = r(s - v),$$
where r is a real scalar. It will be obviously sufficient to prove this for the case in which i and j are unit vectors. Assuming them such, let us write
$$ji = s' + v', \qquad vv' = s'' + v'',$$
where s' and s'' are scalars, while v' and v'' are vectors. Then
$$ij \cdot ji = (s+v)(s'+v') = ss' + sv' + s'v + v'' + s''.$$
But we have
$$ij \cdot ji = ij^2 i = -i^2 = 1.$$
Hence,
$$v'' = 1 - ss' - s'' - sv' - s'v.$$
But v'' is the vector of vv', so that by the last paragraph such an equation cannot subsist unless v'' vanishes. Thus we get
$$0 = 1 - ss' - s'' - sv' - s'v,$$
or
$$sv' = 1 - ss' - s'' - s'v.$$
But a quantity can only be separated in one way into a scalar and a vector part; so that
$$sv' = -s'v.$$
That is,
$$ji = \frac{s'}{s}(s - v). \quad Q.E.D.$$

8. Our next step is to prove that $s = s'$; so that if $ij = s + v$ then $ji = s - v$. It is obviously sufficient to prove this when i and j are unit vectors. Now from any quantity a scalar may be subtracted so as to leave a remainder whose square is a scalar. We do not yet know whether the sum of two vectors is a vector or not (though we do know that it is not a scalar). Let us then take such a sum as $ai + bj$ and suppose x to be the scalar which subtracted from it makes the square of the remainder a scalar. Then, C being a scalar,
$$(-x + ai + bj)^2 = C.$$

But developing the square we have
$$(-x + ai + bj)^2 = x^2 - a^2 - b^2 + abs + abs' - 2axi + 2bxj + ab\left(1 - \frac{s'}{s}\right)v = C;$$
i. e.
$$ab\left(1 - \frac{s'}{s}\right)v = C - x^2 + a^2 + b^2 - abs - abs' + 2axi + 2bxj.$$

But v being the vector of ij, by the last paragraph but one the equation must vanish. Either then $v = 0$ or $1 - \frac{s'}{s} = 0$. But if $v = 0$, $ij = s$, and multiplying into j,
$$-i = sj,$$
which is absurd, i and j being dissimilar. Hence $1 - \frac{s'}{s} = 0$ and
$$ji = s - v. \quad Q.E.D.$$

9. The number of independent vectors in the algebra cannot be two. For the vector of ij is independent of i and j. There may be no vector, and in that case we have the ordinary algebra of reals; or there may be only one vector, and in that case we have the ordinary algebra of imaginaries.

Let i and j be two independent vectors such that
$$ij = s + v.$$
Let us substitute for j
$$j_1 = si + j.$$
Then we have
$$ij_1 = v, \quad j_1 i = -v,$$
$$j_1 v = j_1 i j_1 = -j_1^2 i = i, \quad vj_1 = ij_1^2 = -i,$$
$$iv = i^2 j_1 = -j_1, \quad vi = ij_1 i = -j_1 v^2 = j_1.$$

Thus we have the algebra of real *quaternions*. Suppose we have a fourth unit vector, k, linearly independent of all the others, and let us write
$$j_1 k = s' + v',$$
$$ki = s'' + v''.$$
Let us substitute for k
$$k_1 = s''i + s'j_1 + k,$$
and we get
$$j_1 k_1 = -s''v + v', \quad k_1 j_1 = s''v - v',$$
$$k_1 i = -s'v + v'', \quad ik_1 = s'v - v''.$$

Let us further suppose
$$(ij_1)k_1 = s' + v'''.$$
Then, because ij_1 is a vector,
$$k_1(ij_1) = s''' - v'''.$$
But
$$k_1j_1 = -j_1k_1, \quad k_1i = -ik_1,$$
because both products are vectors.

Hence
$$i.j_1k_1 = -i.k_1j_1 = -ik_1.j_1 = k_1i.j_1 = k_1.ij_1.$$
Hence
$$s''' + v''' = s''' - v'''$$

or $v''' = 0$, and the product of the two unit vectors is a scalar. These vectors cannot, then, be independent, or k cannot be independent of $ij = v$. Thus it is proved that a fourth independent vector is impossible, and that ordinary real algebra, ordinary algebra with imaginaries, and real quaternions are the only associative algebras in which division by finites always yields an unambiguous quotient.

www.ingramcontent.com/pod-product-compliance
Lightning Source LLC
Chambersburg PA
CBHW020059170426
43199CB00009B/333